ISO 14000 Road Map to Registration

ISO 14000 Road Map to Registration

Perry L. Johnson

McGraw-Hill
New York San Francisco Washington, D.C. Auckland Bogotá
Caracas Lisbon London Madrid Mexico City Milan
Montreal New Delhi San Juan Singapore
Sydney Tokyo Toronto

Library of Congress Cataloging-in-Publication Data

Johnson, Perry L. (Perry Lawrence), 1948-
    ISO 14000 road map to registration :   / Perry L. Johnson.
        p.   cm.
    Includes index.
    ISBN 0-07-032910-9 (alk. paper)
    1. ISO 14000 Series Standards.   2. Production management-
-Environmental aspects.   I. Title.
TS155.7.J638   1997
658.4'08—dc21   97-16554
                                                                           CIP

McGraw-Hill

A Division of The **McGraw·Hill** Companies

Copyright © 1997 by The McGraw-Hill Companies, Inc. All rights reserved. Printed in the United States of America. Except as permitted under the United States Copyright Act of 1976, no part of this publication may be reproduced or distributed in any form or by any means, or stored in a data base or retrieval system, without the prior written permission of the publisher.

1 2 3 4 5 6 7 8 9 0   FGR/FGR   9 0 2 1 0 9 8 7

ISBN 0-07-032910-9

The sponsoring editor for this book was Robert Esposito, the editing supervisor was Paul R. Sobel, and the production supervisor was Tina Cameron. It was set in Palatino by Estelita F. Green of McGraw-Hill's Professional Book Group composition unit.

Printed and bound by Quebecor/Fairfield.

McGraw-Hill books are available at special quantity discounts to use as premiums and sales promotions, or for use in corporate training programs. For more information, please write to the Director of Special Sales, McGraw-Hill, 11 West 19th Street, New York, NY 10011. Or contact your local bookstore.

 This book is printed on recycled, acid-free paper containing a minimum of 50% recycled, de-inked fiber.

Contents

Preface   ix
Acknowledgments   xv

1. Crying Over Spilled Oil   1

The *Exxon Valdez* Incident   1
Influences on Environmental Standardization   2
Revised U.S. Coast Guard Shipping Guidelines   2
Potential Benefits of Environmental Management
   Systems   4
Conclusion   5

2. Auditing Begins at Home   9

First-Party (Internal) Audits   9
Advice for Internal Auditors   11
Conformity   12
Third-Party (Registration) Audits   14
Second-Party Audits   20
More on Auditors   22
Evolution of Environmental Management System (EMS)
   Auditing   27
The CERES Principles   33
More on the Role of the Internal Auditor   35
   Conclusion   37

3. Don't Feel Like You're Out of Your "Element" 41
Element 4.1 General   42
Element 4.2 Environmental Policy   42
Element 4.3 Planning   43
Element 4.4 Implementation and Operation   49
Element 4.5 Checking and Corrective Action   55
Element 4.6 Management Review   57

4. Finding a Registrar 59
Credibility of Registrars   59
Auditing Philosophy   60
Steps to Take before Hiring a Registrar   62
Registrar Accrediation   63
On the Horizon: Registrar Membership in Single-Auditor Certification Programs   65

5. The Moment of Truth: ISO 14001 Registration 69
Registration (Also Called Certification)   69
But What Does It Cost?   70

6. SGS-Thomson: Blazing the Trail 73

7. Surveillance Aduits 83
Frequency of Surveillance Audits   83
The Focus of Surveillance Audits   84
The Role of Stakeholder Complaints   84

8. What, You Worry? The Legal Aspects of ISO 14001 Registration 87
Environmental Regulation and the EPA's Role   89
Environmental Regulation and Economic Development—Mutually Exclusive?   95
The People: an Especially "Interested Party"   96
Environmental Activism: Its Own Worst Enemy?   98

9. Suppliers and their Effects on ISO 14000 103
Maintaining Environmental Compliance within the Supplier Chain: A Matter of Common Sense   103
The Virtues of Minor Arm-Twisting   105

Contents   vii

10. Implementation: What It Takes   **107**

The Importance of Training   107
Total Quality Environment Management (TQEM):
  Benefits and Obstacles   116
Greener Grass Inc.: A Hypothetical Case in Point   122
What Are Your Environmental Costs?   133

11. ISO 9000: The Forerunner of ISO 14000   **137**

History of ISO 9000   137
Relationship between ISO 14000 and ISO 9000   139

12. A Sample EMS Manual   **143**

Why an EMS Manual Is Necessary   143
The ISO 9000 Quality Manual as an Example   144
Steps in Writing an EMS Manual   144
How to Start   145
A Sample EMS Manual   147

13. Market Forces and the Future of ISO 14000   **157**

Marketing and Practical Advantages of ISO 14001
  Registration   157
Your New Competitive Edge   158
Parameters of Using ISO 14001 in Advertising   159

14. Conclusion   **165**

ISO 14001 and Social Responsibility   165
Recap of Implementation Steps   167
Scholarly Research in Environmental Management   169
Obstacles to Investment in Preventive Technologies   172

Index   179

Preface

> "It shows our commitment to the corporation's environmental vision to be recognized by all our stakeholders as a leader in environmental care, by exceeding regulatory requirements in both degree and timing wherever possible. It also shows our commitment to the corporation's environmental mission to eliminate or minimize the impact of our processes and products on the environment, maximizing the use of recyclable or reusable materials and adopting, as much as possible, renewable sources of energy, in striving for sustainable development."
>
> JOE HESS, ENVIRONMENTAL ENGINEER
> SGS-Thomson

The need to minimize and repair the environmental damage caused by illegal dumping, fugitive emissions, acid rain, ozone depletion and other forces has become an issue of paramount worldwide importance. What is industry doing to shield the Earth from such potentially devastating impacts?

One effective way of bridging the gap between government regulations and high-cost solutions to these and other problems is ISO 14000.

ISO 14000 is the world's first series of internationally recognized standards for environmental management. The standards represent a coordinated, international approach to global reduction of environmental degradation—one that can be applied to any organization, of any size, anywhere in the world.

The environmental management system standards were created by the International Organization for Standardization—a worldwide federation based in Geneva, Switzerland, comprised of more than 110 member countries. This is the same organization that created the world-renowned ISO 9000 quality-management standards. Today, ISO 9000 is a virtual prerequisite for doing business overseas—particularly in the 15-nation European Union (EU).

The move to create the ISO 14000 standards began in 1991, when the International Organization for Standardization (also known as ISO) created the Strategic Advisory Group on the Environment (SAGE) and charged it with researching the need for international environmental standardization. In less than a year, members of SAGE concluded that there was a definite need for a system of standardized environmental management and, based on this solid recommendation, ISO Technical Committee (TC) 207 was established. TC 207 set about developing an international environmental management system that could be effectively used across international boundaries.

The ISO 14000 consensus documents were scheduled for release in late 1996. But even before the standards were officially released, interest in ISO 14000 was so widespread within the business community that many experts predicted its acceptance would exceed the vast popularity of ISO 9000. Several companies even attained ISO 14001 registration while the standard was still in draft form (one of those companies, SGS-Thomson, is featureed in this book).

The standard's drawing-card is the litany of inherent benefits it provides:

- Improved management of environmental impacts
- Positive public image
- Identified opportunities for continual improvement
- Boosted environmental awareness among employees through training and education
- Reliable management processes
- Removal of environment as defacto trade barriers
- Improved competitive edge in the global marketplace

- Obtaining insurance at a reasonable cost
- Liability limitation
- Satisfying investor criteria
- Cost control
- Showing good-faith effort to comply with governmental regulations
- Ability to dispose of waste effectively and prevent pollution
- Conservation of materials
- Easier site selection and permitting
- Technology development and transfer
- Improved industry-government relations
- Reparation of previously uninhabitable lands, such as brownfields
- And, of course, an enhanced natural environment and higher quality of life

In a world where economic and political concerns continue to dominate the conduct of big business, ISO 14000 offers a new approach to environmental management. It is one that advocates resourcefulness and a positive perception among consumers, and one that can coexist across a wide range of unique national and local laws and regulations.

These developments couldn't have come a moment too soon. According to the book, *Measuring Corporate Environmental Performance: Best Practices for Costing and Managing an Effective Environmental Strategy* (Irwin Professional Publishing, 1996), "environmental management is 'becoming central to corporate strategy and is being mandated as an arena of competition rather than as a compliance-driven function [Lent and Wells, 1993, 379]'."

This change is reportedly being driven by three important and relatively new developments:

- Customers are buying more products that they can identify as having positive environmental attributes;
- The threat of ozone depletion and the response by business to eliminate CFCs brought environmental management into the mainstream of business; and
- Total quality management and pollution prevention are a good fit in that they seek to prevent defects, and thus pollution, before they occur.

We cannot stress enough the competitive advantages of subscribing to the new corporate environmental philosophy. These advantages "are reflected in improved product quality, improved production yields, and improved profitability, the result of redesigned processes and products."

Increasingly, say the authors, "companies have come to recognize that...reduced environmental impacts and the institutionalization of corporate environmental responsibility can lead to improved operations and profitability."

The book says that energy prices have played a prominent role in analyses of environmental initiatives, since low energy *prices* translate into low or no incentive to invest in energy *conservation*. "The social and financial benefits may not outweigh the significant financial costs" of undertaking such seemingly needless conservation, but a bottom-line decision can be made only after a total analysis of stakeholders' needs and expectations. "Stakeholders," in the context of this book, are all parties with an interest in your environmental performance and sense of environmental responsibility.

Energy-producing companies, just one slice of the overall environmental pie, must integrate energy costs—whatever they turn out to be—in the development of corporate strategies in order to better balance the future sustainability of the environment and the corporate well-being.

A thorough understanding of environmental management—the aim of this book—must lead to an overall environmental strategy that serves a springboard to: "set(ting) corporate policies, change corporate culture, and integrate environmental impacts in managerial decisions at all levels, in all facilities, and at all geographic locations of the organization." At the same time, however, the strategy must be wholly adaptable to changing environmental regulations and technologies and be capable of integrating forecasts of such changes into environmental management decision-making.

The result, says the book, must be that of reduced or eliminated waste, reduced environmental impacts and better investments that contribute to long-term corporate profitability.

Implementation of the strategy is ideally carried out through the following corporate conduits:

- Broad-based institutional support for the company's management strategy;
- Development of a strong Environmental Health & Safety organizational structure;

- Environmental policies (an ISO 14000-specific topic we will cover later); and
- Personnel who support and desire increased environmental responsibility and a change in corporate culture.

These tenets are true for small and large companies alike, in both high- and low-impact industries. They must be made a staple of the corporate performance appraisal system, which is an option totally supported by ISO 14000.

Successfully implementing an environmental management system (EMS) based on ISO 14000 is the first step toward realizing all of the advantages the standard has to offer. But registration, sometimes called certification, by an independent third party is the only sure way a company can secure public recognition and acceptance of their management strategies in the future.

The *Road Map to ISO 14000 Registration* will enhance your organization's ability to obtain a certificate of registration. This book will shed light on what you can expect on the journey toward ISO 14000 registration and help you steer clear of the bumps and roadblocks that you may encounter along the way.

Perry L. Johnson

Acknowledgments

I would like to thank several people who helped make this book possible: Howard Gofstein, Molly Dando, and Monica Chapoy Ciagne, for their research, editorial contributions, and review of the manuscript; Andrew Kok, an environmental lawyer from Grand Rapids, Michigan; Paul Fortlage, vice president of operations at the Registrar Accreditation Board (RAB); John Donaldson, vice president for Conformity Assessment at the American National Standards Institute (ANSI); Consultant Jim Scott; Casper van Erp, of the Raad voor Accreditatie (RvA), also known as the Dutch Council for Accreditation; Mary Jane Wilkinson, of the United Kingdom Accreditation Service (UKAS); Environmental Consultants Gene Michael and Ralph Grover; Denise Seipke Environmental Program Manager at Perry Johnson, Inc.; Joel Charm of Allied Signal and TC 207; Joseph Koretsky of electronics manufacturer Sharp; Noel Matthews of the International Auditor Training and Certification Program, Canberra, Austrialia; Ruth Bacon, of the Environmental Auditors Registration Association (EARA); Joanne Moynan, of the Scottish law firm of Shepherd & Wedderburn WS; Richard Webber, of the English law firm of Field Fisher Waterhouse; and Kay Higby of Akzo Nobel Chemical.

SGS-Thomson, Rancho Bernardo, Calif., which graciously shared with us the story of their ISO 14001 registration; and the *Anchorage* (Alaska) *Daily News, for graciously allowing us to use their coverage of the Exxon* Valdez incident.

ISO 14000 Road Map to Registration

1
Crying Over Spilled Oil

The *Exxon Valdez* Incident

March 24, 1989, is a date that few Alaskans will ever forget. It was on that day that the 987-foot tanker *Exxon Valdez* ran aground, turning the icy waters of Alaska's Prince William Sound into a slick, black sea of crude oil.

On that day, and for many weeks afterward, the television airwaves were flooded with images of oil-soaked geese, dead plant life, and blackened shorelines, camouflaged by the tarred bodies of seals, sea otters, harlequin ducks, herring, salmon, and the hundreds of other creatures who fell victim to the worst oil spill in the history of the United States.

According to the *Anchorage* (Alaska) *Daily News,* 11 million gallons of oil poured out of the *Exxon Valdez,* a ship from Exxon Corporation's tanker fleet. It was later revealed that the captain of the vessel had been intoxicated when the ship ran into trouble. In the years since the accident, it has taken more than $3 billion to clean up the coastal waters, and more than $1 billion in civil and criminal settlements to bring closure to the *Exxon Valdez* oil-spill catastrophe.

While it will take many more years for Prince William Sound to fully recover from the long-term ecological effects of the spill, something positive did come out of the incident. The public outcry over the accident prompted the U.S. government to reexamine and strengthen its existing regulations for oil shippers. And perhaps more important, the *Exxon Valdez* spill inspired many U.S. companies to take a closer look at responsible environmental management. As a way of protecting themselves from costly liabilities, companies began to adopt voluntary environmental management programs aimed at pollution prevention.

The adoption of ISO 14000 marks a turning point in pollution prevention, and the curbing of other forms of environmental damage, on a much larger scale. Just as pollution knows no boundaries, the international environmental management system standards were created with no boundaries in their potential application.

ISO 14000 represents an international consensus of nearly all of the world's industrialized nations. In creating the standards, ISO Technical Committee 207 incorporated input from representatives of more than 110 countries.

Influences on Environmental Standardization

Worldwide environmental standards were probably bound to follow other resolutions on international commerce, such as the 1986 General Agreement on Tariffs and Trade (GATT). This controversial accord was intended to set up an international trade framework.

Another influence on standardization may have been the 1992 United Nations Conference on Environment and Development, where delegates set forth 27 principles asserting the importance of preserving a healthy environment for all living things.

The 15-nation European Union (EU) has standards that support its own environmental interests. European industry operates according to the mandatory Eco-Management Audit Scheme (EMAS), which is its model of sound environmental management. It is similar to British Standard (BS) 7750, a voluntary standard that was the precursor to ISO 14000.

Revised U.S. Coast Guard Shipping Guidelines

Because of the multibillion-dollar *Exxon Valdez* incident, the oil industry began to reexamine its environmental management performance and look for ways to make improvements. In the Alaskan region, for example, oil shippers worked together with the U.S. Coast Guard to create several new directives that emphasize responsible and proactive environmental management.

The new directives included

- Documented procedures
- Emergency-preparedness planning
- Extensive training of crew members, including practice drills
- The installation of oil-spill-response apparatus and better monitoring equipment
- The use of escort vessels and navigation specialists

Correspondence of Coast Guard Guidelines with ISO 14001 Elements

Though these directives were drafted before the formation of ISO 14000, it is interesting how closely the guidelines match the requirements of the environmental management system created by ISO 14001.

First, both the oil shippers and the Coast Guard realized the importance of establishing procedures for various activities. Operational procedures specify the way in which an activity is to be performed, including now when, where, why, and by whom. Procedures are vital in ensuring that critical activities are performed consistently and play a large role in ISO 14001, the specification standard to which organizations register (see element 4.4.4 of the standard, "Environmental Management System Documentation").

Second, the new regulations call for emergency-preparedness planning. In order to ensure an appropriate response to possible accidents and spills, operating procedures and controls need to be in place. This is also a requirement of ISO 14001 (element 4.4.7, "Emergency Preparedness and Response").

Third, the new regulations provide for extensive training and practice drills for crew members. The cause of most industry-related accidents can be traced to poor training and human error. To ensure that operators are acting competently, training needs for various tasks must be identified and personnel must receive the appropriate training—and must understand both the benefits of improved performance and the consequences of failing to follow procedures. ISO 14001 requires proper employee training and education in element 4.4.2, "Training, Awareness and Competence."

Fourth, the oil shippers and the Coast Guard identified the need to purchase and/or upgrade critical equipment. The Coast Guard, for example, installed a $7 million system designed to track tanker traffic electronically and added a second person to each shift to be responsible for monitoring the equipment.

ISO 14001 also recognizes the need to provide the appropriate resources and equipment. Element 4.4.1, "Structure and Responsibility," states that "management shall provide resources" (human, physical, and financial) necessary to meet the organization's environmental objectives and targets.

Fifth, each tanker departing from the Valdez Marine Terminal is now followed by a tugboat, which helps steer the tanker if it runs into trouble. Some oil companies provide escort vessels and navigation specialists to help guide oil-laden tankers through areas which are known to be dangerous. This also falls under the aegis of element 4.4.1.

Most notably, the voluntary directives created by the oil shippers and the Coast Guard are aimed at pollution prevention, an underlying principle of the ISO 14000 series. The standards help companies address their environmental aspects consistently and in an orderly fashion through the implementation of an environmental management system (EMS) (or the improvement of an existing EMS), the allocation of resources, the assignment of responsibility, and the continuous evaluation of practices, procedures, and processes.

Potential Benefits of Environmental Management Systems

When an effective EMS is in place, an organization can greatly reduce the negative environmental impacts of its activities, products, and services and help maintain the quality of the ecosystem. An organization which has successfully implemented an EMS can also achieve significant competitive advantages: having a "good corporate citizen" image in the community, enhanced market share, improved management of regulatory compliance, the ability to obtain insurance at a reasonable cost, and much more. All it takes is commitment, good planning, and the provision of necessary resources.

Akzo Nobel's ISO 14001 Registration

This kind of commitment was practically incidental for Akzo Nobel Chemical's facility in Axis, Alabama, which had been operating under a comprehensive environmental management system since 1990.

Responsible Care superintendent Kay Higby said the facility's environmental commitment had been so strong that it was just a matter of time before it was formalized in some way—if not through ISO 14000, then by some other means.

Akzo Nobel's corporate office in the Netherlands was always behind the decision and supported it unequivocally, Higby said.

"We felt it was the right thing to do, considering the environmental commitment we have at this time," she said, adding that there had been no significant environmental impacts associated with the facility. The site, north of Mobile, manufactures chemicals used in other manufacturing processes. It sells sulfuric acid to paper mills and carbon dioxide to rayon manufacturers, among others.

The only areas specified by ISO 14000 but not covered by the existing EMS were aspect identification, document control, and auditing. Because most areas were covered, however, and the system bore a strong resemblance not only to ISO 14000 but to ISO 9000, registration took only about six months.

"Our registration is so based on the existing site system that there were no major obstacles. We were pleased to see how well they matched. We felt that we would be able to slightly improve the system because we had already made a global effort," Higby said, adding that registration has kept the company's reputation strong and positive.

She noted that one selling point of ISO 14000 was its emphasis on prevention, rather than curing, of environmental ills.

Parallels between the facility's existing EMS and the two standards, ISO 14000 and ISO 9000, motivated a try for dual registration. Following a single joint audit, the facility's registrar, Bureau Veritas Quality International Ltd., granted registration to ISO 14000 and 9000 simultaneously.

"We did feel it would improve our relationship with the community," Higby said. "We've received positive recognition and we're happy about that."

Conclusion

Because of annual settlement payments scheduled to run through the year 2001, Exxon Corporation continues to be reminded of its carelessness and the consequences that can result when business activities negatively affect the environment. The *Exxon Valdez* oil spill blackened more than just the shores of Prince William Sound on that fateful day in March of 1989, but the public can take comfort in knowing that more

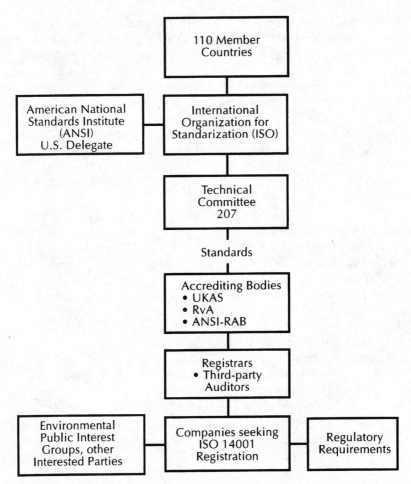

Figure 1-1. Companies seeking registration to ISO 14001 are influenced, both directly and indirectly, by the actions of accrediting bodies, regulatory agencies, and interested members of the general public.

and more companies now realize the importance of responsible environmental management and are taking action to identify activities that can interact with the environment, control activities which involve those aspects, and minimize pollution.

ISO 14001 is a dynamic, highly adaptable tool that organizations can use to establish a general sense of direction in their efforts to improve environmental performance while integrating environmental considerations into all business decisions. The time has come for all companies to act responsibly, and ISO 14001 may be the best solution for bridging the gap between government command-and-control tactics and good business sense.

Any company seeking ISO 14000 registration is influenced by external stakeholders ranging from the official bodies who accredit registrars (which in turn grant or deny registration) to the legislative requirements and special interest groups which drive environmental compliance and responsibility. Figure 1-1 illustrates this hierarchy.

Throughout the rest of this book, we will discuss these parties and their roles as we concentrate on ISO 14001 registration and walk you through each step in the process.

Ready? Let's get started.

2
Auditing Begins at Home

Auditing is the method by which we determine whether an environmental management system is in conformance with ISO 14001. In this chapter, we will walk chronologically through all the audit types in the order in which they occur, explaining how and why each is done. Later in the book, we'll cover the details of third-party audits, including timetables and costs.

First-Party (Internal) Audits

Suppose you are an executive in a company which has just implemented an EMS in accordance with ISO 14001. Presumably, you long ago sent at least a few of your employees to an EMS internal auditor course; these people will now put what they learned into action.

Your company must undergo internal audits well in advance of your registration (third-party) audit to confirm that you are following ISO 14000's mandates, to identify needs for change and to initiate whatever corrective action you have deemed necessary.

Internal audits are also held at regular intervals following registration, in order to confirm that your EMS continues to be effective. First-party audits can be done by your company's staff or by contracting with an outside party. You are free to conduct this audit in whatever manner fits your needs, as long as you cover all the bases. First-party auditing is the least formal type of auditing.

Again, an audit is an evaluation of an implemented EMS in order to assess whether it meets the requirements of the ISO 14001 standard. If your employees took an internal-auditing course, they learned that audits typically have a life cycle which includes writing an audit plan; holding an opening meeting with managers of the area being audited, at which agreement on the audit plan (time and location of audit) is formalized; performing the actual audit, gathering objective evidence through interviews and observations; recording findings; and reporting those findings to the auditee at a closing meeting.

A fully implemented EMS may have an effect on the operations of several areas within your company. It may be more convenient, however, for your company's internal auditors (hereafter referred to as the *audit team*) to audit your EMS piecemeal, that is, one or two areas at a time.

The process needn't proceed with the formality of a registration audit, which by definition is structured and rigid. Remember, all you are attempting to do at this point is to satisfy yourself that you are ready for the registration audit. There are no rules, aside from those regarding auditing in general, to tell you how you must proceed within your own company.

The opening meeting in a first-party audit need consist only of the audit team's telling management, with representatives of the areas to be audited sitting in, "We're going to go to the training records file cabinet and make sure the records are in order" or "We're going to make sure that the company is staying on top of relevant regulatory requirements."

The audit team should be prepared to record its findings as it visits the audit site. In order to ensure objectivity, no person who regularly works in a given area should audit that area. The team is seeking to collect *objective evidence* as to whether the site it is auditing conforms to the applicable standard.

Objective evidence of this nature is collected by the following means:

- Interviewing individuals involved in activities related to the EMS. This is the most effective auditing technique, because it allows you to explore a number of issues face to face as certain questions lead to more questions.
- Examination of EMS documents, such as training records and records of emissions.
- Observation of activities, another form of objective evidence. An auditor may do this by simply standing quietly in the background, out of everyone's way, watching operations in action and taking notes.
- Generally inspecting procedures, products, etc.

Advice for Internal Auditors

When interviewing, remember that how you ask questions determines in large part the answers you'll receive. It is essential, out of respect for the interviewed party's time, that you plan ahead. Decide the scope of your questions and the amount of time needed. Remember that the people you will interview are unlikely to be as familiar as you are with the reason for the audit or the nature of management system models.

Keep your questions simple. Don't just rephrase, in question form, an element of the standard against which you're auditing. Instead, create an atmosphere of trust and communication, if you expect to obtain the information you need.

Talk to a machine operator or clerk with the intention of finding out what it is he or she does, not whether he or she is doing it correctly. Speak the person's language. Show respect for the person's level of knowledge and expertise concerning his or her job function. Stress also that you're there to audit the environmental management system, not the person's performance.

In order to get the most information possible, ask open-ended questions. Give an operator the opportunity to explain as much as possible about his or her job. Questions starting with words like *who, what, when, where,* and *how* are likely to garner much more information than questions that can be answered with a simple yes or no.

Asking an operator to demonstrate her or his job for you, showing a genuine interest and curiosity on your part, will elicit a wealth of information and may even prompt additional questions. People love to talk about their work. If you're sincerely interested and it shows, you will have established exactly the kind of rapport needed to obtain the information you want.

One last piece of advice on interviewing during an audit: Question the person who actually changes the filter, files the documents, conducts the training, or performs whatever other task may be involved. Don't undermine anyone, or your own credibility, by going above his or her head and talking to a supervisor. A supervisor is not likely to be as good a source of information as the person who has firsthand contact with a particular task on a daily basis.

Employees chosen to operate as internal auditors should be of a professional demeanor, with excellent communication skills, able to understand processes and think critically. They should not

- Be sarcastic
- Swear or curse

- Display negativity
- Criticize
- Get sidetracked
- Argue
- Compare employees to one another
- Discuss company policies
- Discuss controversial topics, such as politics, religion, or sex
- Drink before an audit task
- Be late

Conformity

One of four conclusions will be arrived at during the audit: "conformity," "major nonconformity," "minor nonconformity," or an "observation." Let's define these terms now.

- *Conformity:* The state of being in conformance with the requirements of the standard.
- *Major nonconformity:* The absence of a required procedure, so that an element of the standard is not being addressed, or the total breakdown of a procedure, which can cause a systemic breakdown within the organization's environmental performance.
- *Minor nonconformity:* A single, isolated, observable, and easily rectified lapse in a procedure which does not pose a serious threat to environmental performance.
- *Observation:* A situation which is not yet a nonconformity, but may lead to one.

When the auditors are evaluating their designated sites, they will indicate any nonconformity on a nonconformity report form, also known as an NCR. A representative of management should accompany the internal auditors and answer any questions; this representative should add his or her signature to the NCR, indicating his or her awareness of the situation.

A separate NCR should be filled out for each nonconformity and should include

- The audit date
- The area in which the nonconformity was discovered
- The standard against which the audit was conducted

- The ISO 14001 clause against which the nonconformity was found
- Acknowledgment of management awareness of the nonconformity
- A narrative—a brief description of the nonconformity
- The auditor's name

According to the standards by which trained auditors operate, a thorough evaluation of a nonconformity consists of

- Collecting all available evidence
- Making precise written observations
- Establishing the facts
- Discussing and verifying concerns
- Notifying the management liaison, who then signs the NCR, thus acknowledging the need for corrective action

When any nonconformities have been evaluated, the audit of that section of the company is then complete. Then, at your earliest convenience, another portion of the company may be audited, and so on, until a complete internal audit of your facility has been performed.

The closing meeting of an internal audit, like the opening meeting, is brief. No formal audit report need be written; the management liaison to the audit takes the NCRs, and all parties agree on deadlines for fixing (taking corrective action on) the nonconformities.

Because internal audits are informal, cooperative efforts by personnel within a given company, audit findings, and possible solutions are openly discussed among management, employees of the audited area, and members of the audit team. In a registration audit, as we shall see, information isn't shared nearly as openly, and the third-party audit team is not permitted to make recommendations regarding corrective action. Third-party auditors are merely visiting delegates of a company called a *registrar*. And for a registrar's auditors to do any consulting, or to offer to do so, represents a major conflict of interest.

Internal-auditor training may be sufficient to qualify employees to perform audits within their own company. Individuals who have environmental education and experience are preferable to those who have quality auditing experience. Special training will probably be needed to assist individuals who do not have environmental background. The people within your organization whom you designate as internal auditors, however, should also possess certain personal attributes that will help them navigate the waters of EMS auditing. They should be

- Diplomatic

- Professional
- Articulate
- Judicious
- Communicative
- Honest
- Unbiased
- Inquiring
- Observant
- Understanding
- Industrious
- Fair-minded
- Thick-skinned

The auditor must understand that an audit is not a search-and-destroy expedition. He or she is evaluating a process, not the operators of that process. But will his or her coworkers—the ones being asked probing questions—also understand this? The fact that auditor and auditee are already acquainted contributes to the informality of the internal audit process. It does not, however, detract from the objective precision with which that audit must be performed.

Following the institution of all necessary corrective action, your EMS should operate for between three and six months before you schedule your registration audit. The reason for this is that you must accumulate an archive of EMS records for evaluation as part of your registration audit. Let's look at that audit next, followed by a discussion of second-party auditing.

Third-Party (Registration) Audit

The registration audit process determines whether you will become registered to ISO 14001. It is a formal and serious undertaking, requiring careful coordination between your facility and the registrar.

The registrar will ask you to fill out a questionnaire, on which you will indicate certain information about your company: physical size of your facility, range of the company's products and services, level of automation, number of processes and services, how many employees you have, how many shifts you run, and other background. The registrar will use this information in determining how many auditors it must send to your facility for a preregistration audit (also called a preassessment), document review, registration audit, and surveillance audit.

You will receive a proposal of times and costs for all these activities. Let's look at them one by one.

Preassessment

A preassessment is an audit in which the registrar evaluates your EMS in order to point out nonconformities that you must correct before you can pass a registration audit. Although it is an integral part of the overall registration process, the preassessment is not something you can pass or fail. It is performed entirely for your benefit. Think of it as a dress-rehearsal audit.

The registrar will submit its findings in writing, giving you a detailed map of all outstanding nonconformities.

Along with this report, you will receive a formal proposal for the registration audit.

Document Review

Prior to the registration audit, the registrar performs a document review, also called a "desktop audit," in order to determine whether your EMS documentation indicates that you have established, implemented, and maintained an EMS in accordance with the requirements of ISO 14001. The document review is mandatory; where it is performed is entirely up to the manager of the facility being audited.

It is less expensive to ship all of your EMS documentation to your registrar so that the auditors can review it at their office (thus the term *desktop*). If you choose, you can ask the registrar to conduct the review at your facility, but this will involve travel costs and the need to provide a work space and all necessary communication and clerical support for the auditor(s).

The document review can reveal whether you have addressed all the requirements of ISO 14001. As a cross-reference of ISO 14001 against your facility's documented practices, the registrar can also develop a registration audit checklist and finalize its registration audit schedule for your facility. You will receive this schedule, complete with the registrar's itinerary and other written arrangements.

Registration Audit

The registration audit can be likened to a visit from a head of state. The entire visit is very formal, very structured, and scheduled to the minute.

Let's look at how a registrar prepares for a registration audit.

The registrar proceeds according to a plan. Some of the things included are

- A commitment to be flexible (the plan is only a guide—other questions are sure to arise)
- When and where the audit will occur
- Audit scope and objectives
- Identification of key auditee personnel
- Identification of the most important aspects of the EMS being audited
- Expected time and duration of major audit steps
- Procedures for actually conducting the audit
- References to the applicable standard or standards
- Identification of audit team members
- Any meetings that are scheduled
- Distribution list for the final audit report
- The report's planned format
- Agreed-upon requirements for retention of pertinent documents

Many of the items in the audit plan may seem obvious. But a registrar is unlikely to leave anything to chance. And because this plan has resulted from discussions between you and the auditor, you probably have also made arrangements for things such as availability of office space and fax and telephone service, a guide who will walk the auditors through the facility, and safety provisions (Are there certain areas in which auditors should wear protective gear? Is that gear available?).

At the opening meeting between your facility's representatives and third-party auditors, including the team's lead auditor, you will have your first encounter with the individuals who will evaluate your EMS and reach a decision about whether or not you have earned ISO 14001 registration.

Who are these people? What qualifies them as EMS auditors?

Auditor Qualifications

Two ISO 14000 documents, ISO 14010 and ISO 14011/1, give guidelines for EMS auditing and auditor qualifications. ISO 14012 lays down requirements against which third-party EMS auditors and lead auditors must be certified.

EMS auditors, according to ISO 14012, should have completed at least their secondary education or the equivalent. Those who have earned a college degree must have four years of appropriate work experience. Prospective auditors who have completed only their secondary education should have five years of experience. This work experience may be in any of the following disciplines:

- Environmental science and technology
- Technical and environmental aspects of facility operations
- Relevant requirements of environmental laws, regulations, and related documents
- Environmental management systems and standards
- Audit processes, procedures, and techniques

On-the-job training should occur in the same areas. Auditors' specific responsibilities are

- Following and supporting the directions of the lead auditor
- Planning and carrying out the audit in accordance with applicable auditing requirements and other directives
- Complying with and communicating audit requirements
- Collecting and analyzing evidence in order to draw conclusions
- Preparing working documents that will be used to collect evidence
- Remaining within the scope of the audit and exercising objectivity
- Reporting critical nonconformities to the lead auditor
- Reporting major hindrances interfering with the audit
- Maintaining ethical conduct
- Reporting audit results clearly and assisting in the writing of the audit report

An EMS auditor or lead auditor candidate should have completed 20 equivalent working days and four EMS audits under the supervision of a lead auditor. These audits should encompass the entire audit process and occur within not more than three consecutive years.

The lead auditor will be your primary contact during the audit because he or she supervises the entire audit team and oversees the process. He or she is the spokesperson for the team during both the opening and closing meetings and is responsible for

- Selecting the audit team members
- Making administrative decisions
- Overall management of the audit team
- Working with the client to determine the scope of the audit
- Preparing the audit plan
- Coordinating the preparation of working documents
- Serving as liaison
- Reporting major obstacles encountered during the audit
- Conveying safety-related audit issues to the audit team
- Immediately reporting critical nonconformities to the party being audited
- Submitting the audit report containing the audit findings

In order to become a lead auditor, an individual must be able to prove, through interviews, references, and other means, that he or she has the strong management and leadership skills needed to oversee the audit process. Lead auditor candidates must be able to prove that they have completed at least three audits, while acting as lead auditor during at least one of those three audits. All of this must have taken place within a maximum of three consecutive years.

It's not easy. At least one major accrediting body, the United Kingdom Accreditation Service (UKAS), requires lead auditors to possess the ability to audit not only against ISO 14001, but against the far more stringent European Eco-Management Audit Scheme (EMAS) and the British BS 7750, two standards that are the basis of the ISO 14000 series. Because no one could immediately comply with every one of *these* EMS lead auditor criteria, a *transitional* arrangement was established and scheduled for review in 1997. Under this arrangement, a person could be certified as an EMS lead auditor under the following conditions:

- Qualification in the scope or sector of the organization being audited
- Having acted as lead auditor in two out of four EMS audits in which he or she has participated
- Sufficient EMS training, including attendance at an EMAS course
- Familiarity, through training or experience, with environmental legislation and national regulations in the country in which the audit is taking place

Nonconformities and Audit Reports

We mentioned the types of nonconformities your internal auditors may find. When a nonconformity is found during an internal audit, it has no greater repercussion than creating the need for corrective action. Nonconformities will be far more serious, however, if they are allowed to linger until the third-party registration audit. They will delay your registration. In more extreme cases, major nonconformities can

- Have a negative impact on the environment
- Greatly affect an organization's environmental goals and objectives
- Jeopardize compliance with applicable laws and regulations
- Expose the organization to legal liability
- Cause great harm to other environmental operations within the company

Here are some examples of major nonconformities:

- An absence of documented procedures for communication with internal and external stakeholders
- Improper or inadequate identification of significant environmental aspects or impacts
- No reports of corrective action or other follow-up to previous audits
- Uncalibrated measuring equipment
- No record of changes to plans and/or drawings

And a few minors:

- Isolated instances of documentation that denote unauthorized changes
- Isolated examples of deficient record keeping
- Inadequate documentation of training requirements

Be they major or minor, nonconformities discovered by third-party auditors may delay a company's certification to the ISO 14001 standard. All nonconformities must be corrected, or, to use EMS jargon, "closed out," by the registrar or other certifying body before certification can occur.

Like NCRs in internal audits, NCRs in third-party audits should be signed by both the auditor and the auditee representative, confirming that the auditee is aware of the nonconformity. This avoids surprises at the closing meeting.

After audit activities are concluded, a closing meeting is held. At this meeting, the lead auditor summarizes the team's findings, individual auditors may choose to offer highlights of their experiences, and all checklists and NCRs are given to the auditee, along with a short report documenting the team's major findings.

Later, an official audit report is delivered to the auditee. The distribution list for this report is decided between the two parties at the closing meeting. The report may be written solely by the lead auditor, or it may be a collaborative effort by the entire audit team. It is confidential and contains

- Identification of the auditee's representatives who have been privy to the audit process
- Identification of the audit team members
- The audit dates
- A restatement of the agreed-upon audit scope
- The standards against which the audit was conducted
- A recap of how the audit went, including obstacles, if any, that were encountered (a rarity)
- A listing of the reference documents that were used
- Nonconformity reports, differentiating between major and minor
- Conclusions with respect to the portions of the EMS that were audited, including a judgment about the extent to which the EMS currently conforms to the standard and the system's ability to meet its stated environmental objectives

The registration audit report will not contain recommendations for correcting nonconformities or offers of consulting services of any kind. Remember, for a third-party auditor to recommend corrective action to you is a conflict of interest.

Deadlines for follow-up action will be agreed upon by the two parties. Depending upon the nature of the nonconformity or nonconformities in question, the registrar may require that a revised version of the relevant documentation be mailed to its office.

Second-Party Audits

A second-party audit occurs when a customer evaluates one of its suppliers against applicable elements of the ISO 14001 standard. A company that is registered to ISO 14001 needs to know that the company

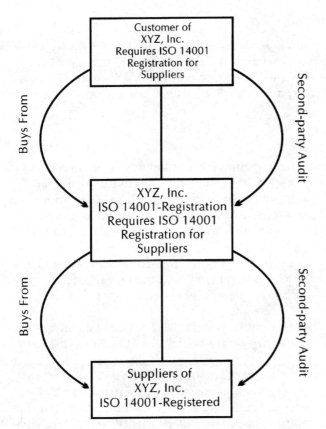

Figure 2-1. In the second-party audit, companies seek to verify that their suppliers are in compliance with ISO 14001—or, at the very least, that they employ common-sense principles of environmental management.

from which it obtains raw materials and equipment, or the firm which disposes of its hazardous waste, isn't going to throw it out of the state of conformance it has worked so hard to achieve. Figure 2-1 illustrates the second-party audit.

Second-party EMS auditing is likely to become more pervasive as ISO 14001 registration catches on. At present, companies aiming for ISO 14001 registration are, justifiably, more concerned about getting their own ducks in a row.

More on Auditors

At least two accrediting agencies—specifically, the Environmental Auditors Registration Association (EARA) and the Dutch Council for Accreditation (RvA)—require that prospective EMS auditors prove themselves beyond the contents of ISO 14012.

EARA

EARA is an independent, nonprofit organization whose mission is raising the standards of professional competence for environmental auditors. Gaining a spot on its published register of qualified EMS auditors, which contains detailed information about each registrant, requires that you meet a set of requirements which far exceed those of the ISO 14000 series.

EARA classifies its registered auditors into three categories—associate environmental auditor, environmental auditor, and principal environmental auditor—depending upon their qualifications and demonstrated abilities. Points are assigned according to how well one meets the various criteria. The criteria for associate environmental auditor are the least stringent; those for principal environmental auditors, the most stringent.

Environmental auditors and principal environmental auditors must possess the types of interpersonal skills we have already mentioned: competence in expressing concepts and ideas; diplomacy, tact, and the ability to listen; the ability to maintain independence; and objectivity, organization, and sound judgment.

All applicants should become competent in EMS auditing through either formal or on-the-job training lasting no less than five days. Training should occur in the following areas:

- Environmental management systems and standards against which audits may be performed (ISO 14001, EMAS, and/or BS 7750)
- Environmental processes and effects
- Requirements of environmental laws and regulations, and related documents
- Audit procedures, processes, and techniques
- Techniques of environmental monitoring, control, and mitigation

Associate auditor candidates without this training are required to attend EARA's foundation course in environmental auditing before they can appear in the EARA register. EARA encourages even those candidates who have the prerequisites to enroll in its training courses.

Background experience for those applying for the environmental auditor and principal environmental auditor levels should have work experience in these areas, according to EARA:

- Environmental science and technology
- Technical and environmental aspects of facility operations
- Relevant requirements of environmental laws
- Environmental management systems and standards
- Audit procedures, processes, and techniques

An applicant who has a university degree in engineering, natural sciences, business management, or law should have at least two years' work experience. An applicant who has a degree in another field, or who has two or more higher education qualifications, needs a minimum of three years' work experience.

Applicants who do not possess a university degree but have only a secondary education need four years' work experience. In exceptional cases only, EARA says, an individual can reduce the required work experience by one year if he or she has completed full- or part-time education in the disciplines listed above.

Beyond these requirements, EARA mandates an interview for those prospective environmental auditor candidates who are considered borderline in their qualifications. All principal auditor candidates must submit to an interview, and also provide a 5,000-word thesis on a topic pertaining to their auditing experience.

Here are the registration requirements specific to each auditor level (taking a lead auditor course such as that offered by Perry Johnson, Inc., the largest trainer of lead auditors in North America, is a good foundation upon which to build competence as an auditor).

Associate Environmental Auditor. Associate auditor candidates should have at least a basic knowledge of EMS auditing and be actively engaged in auditor training. They should be able to accumulate at least 10 points based on the following:

Academic Qualifications. Up to 5 points.

Applicants must have at least a secondary education, although this alone is not sufficient to earn an applicant any points. Points are gained by possessing relevant academic experience in any engineering field, the natural sciences, business management, or law. A bachelor's or similar-level degree is worth 2 points; a higher degree, such as a master's or doctorate, is worth an additional 3 points if it is earned in a field unrelated to the bachelor's degree. If the field is the same, only 1 additional point is awarded.

Degrees that are less relevant to environmental study—language, history, and social science—are worth only 1 point, with additional secondary educational qualifications (such as technical certificates) being worth ½ point.

Professional Membership. Up to 5 points.

Full membership in some trade associations is worth between 2 and 3 points, and fellowships in these bodies are worth up to 4 points. Points are not awarded if membership was achieved by nomination or academic qualifications.

Training. Five points maximum.

Anyone who has less than 25 days of auditing experience is required to attend an EARA-accredited course. But all applicants to this level are encouraged to undergo such training regardless of their experience.

How many points are earned through training depends upon the applicant's performance throughout the course, as well as his or her final grade. An applicant is awarded 5 points for a final score of 70 percent; 3 points for a score between 55 and 70 percent, and 1 point for a score between 45 and 55 percent. No points are given for scores below the 45 percent mark.

Grades from EARA-approved courses *only* will be considered for associate auditor–level candidates. A list of these courses is available from EARA.

Experience. Number of points depends upon experience.

Applicants for the associate level are required to earn 2 points from work experience.

1. Development, implementation, auditing, and assessment of EMSs: 1 point per 5 days' experience

2. Environmental auditing, other than EMS auditing: 1 point per 5 days' experience

3. Other auditing experience, including participation in second- and third-party audits: 1 point per 10 days' experience, with 3 points maximum.

4. Other appropriate work experience in some or all of the following: environmental science and technology; technical and environmental aspects of facility operation; or relevant requirements of environmental laws, regulations, general environmental management stan-

dards, or nonenvironmental audit procedures, processes, and techniques, 1 point per 20 days' experience.

Environmental Auditor. Applicants for environmental auditor must possess at least 100 days of environmental auditing experience, including four supervised audits taking 20 days, in addition to the prerequisites listed for associate environmental auditors. In all, their background must be worth at least 30 points.

Environmental auditor candidates must also be competent auditors with a good appreciation of auditing techniques and environmental issues, even if their experience is confined to only a few audit types and sectors. They may occasionally be selected to undergo an interview.

Principal Environmental Auditor. Principal environmental auditors must earn 50 points and have at least 200 days of environmental auditing experience. One hundred of these days must consist of "hands-on," in-the-trenches EMS auditing, comprising a combination of multi-issue compliance audits, single-issue audits, environmental risk assessment, life-cycle assessment, and EMS internal audits.

They should have plentiful auditing expertise and the ability to lead and manage audit teams. They are the EARA equivalent of lead auditors.

Applicants here must submit a 5,000-word treatise and be tested on it by a peer group. They must have served as lead auditor through at least seven audits in a 35-day period.

The thesis should cover a topic within the candidate's own practical experience as an environmental auditor. It must not, however, be highly theoretical or overly simplistic. If they choose, candidates for principal environmental auditor registration may submit a completed audit report if they had primary responsibility for leading the audit and for writing the report.

All applicants must pass a rigorous assessment that includes references, background checks, and peer reviews. They must agree to abide by EARA's 12-point code of ethics.

Once registered, auditors are expected to maintain their competence by the following means:

- Keeping their knowledge of EMS standards current
- Ensuring that their knowledge of environmental laws and regulations and audit processes is current
- Taking refresher training as needed

- Keeping their experience current

This continuing education must be documented in a log book, which is reviewed by employers and clients, and submitted annually, and on demand, to the EARA Secretariat. The information can also be used as a basis for upgrading registrants' status, such as from associate environmental auditor to environmental auditor.

Registration for higher-level auditor classifications must include the following experience:

- Internal EMS auditing
- Audit for BS 7750 certification
- Verification of company audit findings or environmental audit statements
- Multi-issue site audits against corporate environmental policies and/or environmental legislation
- Single-issue (e.g., waste minimization) audits
- "Green" product audits; life-cycle analyses
- Risk assessments, environmental liability, or due-diligence audits for companies involved in divestiture or merger activities

RvA

The RvA—the Dutch Council for Accreditation, also known as Raad voor Accreditatie—expects prospective EMS auditors to have up-to-date work experience and training in these areas:

- Techniques to reduce harmful environmental impacts and the application of these techniques in practice
- Performing environmental effects analysis
- Relevant regulatory requirements

The council mandates that audit teams consist of at least one lead EMS auditor and that all team members be familiar with

- The general concept of management systems
- The most important effects on environmental media
- Auditing principles

Not all team members need to be experts in these areas; however, at least one person on the team must be certified to lead an audit on every applicable element of an EMS. Also, an audit team may consist of only one person if necessary, and a team may include experts with specific knowledge about environmental issues specific to the company being audited, provided that they are not assigned independent auditing activities.

EMS auditors should have a level of knowledge in an environmentally relevant field beyond the secondary education level. If an auditor's education does not exceed the secondary level, relevant work experience can make up the difference.

Auditors should have at least four years of such experience in either case. This experience should expand their knowledge of these areas (especially if they are acting as lead auditors):

- Regulatory and legal compliance in the environmental field
- Management systems and auditing methods
- Mitigation and control of environmental impacts and aspects
- Sector-specific knowledge pertinent to the audit

As well as meeting the above requirements, the lead auditor must be able to demonstrate a thorough understanding of how to apply positive personal attributes and skills to the task of planning, controlling, managing, and recording an effective audit.

Evolution of Environmental Management System (EMS) Auditing

The concept of environmental auditing was not born concomitantly with the development of ISO 14000. Several types of environmental auditing exist and have been in use for years, with EMS auditing being the newest addition to the list.

If we accept the notion of auditing as an integral part of any environmental management system—and we stipulate that such audits are a response to *social* pressure to control environmental impacts and obey applicable laws—then the first true environmental audits took place as far back as the 1970s.

In "Attributes of the Environmental Management Systems of Manufacturing Firms: The Role of the Internal Auditor in Environmental

Auditing,"* Sharon Nelms Campbell wrote that "during the 1970s, as a result of societal pressure, some corporations attempted to conduct *social audits*" (p. 40, emphasis added). So even though an audit can be a key component of a comprehensive environmental management system (p. 40), the concept seems to have begun as an attempt by corporations to please outside stakeholders from all sectors of society.

Campbell writes that the more environmentally intensive industries were the first to conduct such reviews. "While many entities, including individuals, households, government units, and service industries have contributed to environmental problems, the environmental effects of the activities of the manufacturing and extractive industries were often more dramatic or more readily apparent. Thus, early social audits often focused on the assessment of pollutant emissions by corporate factories, mines, and products, particularly air pollutants" (p. 41).

The truth is that before the advent of ISO 14000 and its required EMS audits, *environmental* auditing was and still may be used for a "wide range of management practices which are generally undertaken on a voluntary basis by companies as part of a strategy of self-regulation" (p. 7).

These activities, even in a non-ISO 14000–registered facility, can range from legal compliance to assessing the adequacy of the environmental management system in use. In either case, in order for development to be sustainable, both environmental and economic concerns must become part of the decision-making processes of organizations—something we have tried to facilitate by writing this book.

Campbell studied organizational responsibility for environmental auditing, types of environmental audits conducted, sources of personnel (internal or external) who conduct the environmental audits, and the benefits and risks of environmental audits—topics we shall cover shortly. She discovered that auditing is greatly encouraged by regulatory bodies like the Securities and Exchange Commission (SEC) and, of course, the Environmental Protection Agency (p. 41).

The SEC's involvement stems from two laws which say that publicly held companies have a legal obligation to disclose environmental (management) information if it is material to shareholders and investors. Other regulations mandate release of information about capital expenditures for environmental control facilities (p. 42).

---

*Sharon Nelms Campbell, "Attributes of the Environmental Management Systems of Manufacturing Firms: The Role of the Internal Auditor in Environmental Auditing," doctoral dissertation, Louisiana Tech University, 1995.

The EPA's interest in all things environmental, especially auditing, stemmed from the proliferation (Campbell, p. 43) of environmental regulations during the 1970s and into the 1980s. "Environmentally intensive" (p. 43) industries saw auditing as a means of evaluating their compliance with these regulations. The EPA itself recognized environmental auditing as a means to fulfilling companies' private and public goals, and went so far as to incorporate auditing in a 1986 policy statement (p. 43). At the same time, however, the agency said it understood the need to preserve the privacy of audit results (p. 44). Officials acknowledged that some firms might not want to implement "rigorous" auditing programs if public disclosure were required:

> EPA believes routine Agency requests for audit reports could inhibit auditing in the long run, decreasing both the quantity and quality of audits conducted. Therefore, as a matter of policy, EPA will not routinely request environmental audit reports. EPA's authority to request an audit report, or relevant portions thereof, will be exercised on a case-by-case basis where the Agency determines it is needed to accomplish a statutory mission, or where the government deems it to be material to a criminal investigation (p. 44).

An Environmental Protection Agency news release dated March 31, 1995, indicates, however, that the EPA will "substantially reduce" penalties on facilities which "voluntarily disclose and promptly correct" violations which are identified through "self-policing and meet certain other specific conditions." Under this policy, initiated by the Clinton administration, the Justice Department would be discouraged from prosecuting such facilities.

"This new policy is a good example of the Clinton administration's common-sense approach to environmental protection," EPA Administrator Carol Browner is quoted as saying. "It is balanced and carefully crafted to give the regulated community predictability and to reward environmentally responsible actions."

The EPA's policy provides for the requesting of reports only when other, independent evidence exists that a violation has occurred. Steven Herman, the EPA's assistant administrator for enforcement and compliance assurance, said, "We are not going to use audits to engage in fishing expeditions." The policy was scheduled to take effect in mid-April, and copies are available by faxing (202) 260-7548.

The state of Michigan in 1996 added a similar provision to its Natural Resources and Environmental Protections Act. The amendment shields some environmental audit information from public disclosure, giving facilities an incentive to perform audits, while continuing to grant to the state's Department of Environmental Quality the authority to access any information necessary for the enforcement of state laws.

Other so-called self-audit legislation has been passed (as of May 1996) in Arkansas, Colorado, Idaho, Indiana, Illinois, Kansas, Kentucky, Minnesota, Mississippi, Oregon, Texas, Utah, Virginia, and Wyoming.

Campbell found that audits are conducted for three main reasons: compliance with external regulations and internal policies, identification of potential environmental liabilities, and evaluation of the environmental performance of other parties (p. iv). While she was able to find little empirical evidence on the overall involvement of the internal audit function in environmental auditing, she did determine that there are several types of audits that may be used to achieve these three objectives. Which are used, and to what extent, varies from organization to organization. These variations depend upon management's bottom-line environmental objectives (pp. 48–54).

Audit of Compliance with Environmental Regulations

The very minimum goal of environmental auditing is to assess whether a facility is operating within its legal boundaries. Environmental regulations are imposed at all levels of government (federal, state, and local) and are designed to protect the key environmental elements: air, land, and water. There are several major federal laws, such as the Clean Air and Water Acts, Toxic Substances Control Act, Emergency Planning and Community Right-to-Know Act, Resource Conservation and Recovery Act, and CERCLA (Comprehensive Environmental Response, Compensation and Liability Act). CERCLA is commonly referred to as "Superfund."

As well as determining compliance with applicable laws, audits may also be used to provide information on what is necessary to achieve compliance and what it will cost. So-called compliance audits, like ISO 14001 EMS audits, range from a simple document review to a full-blown site visit.

Audit of Compliance with Company Environmental Policies and Procedures

Some, if not most, facilities have policies unique to them which are used to control environmental impacts and assure regulatory and legal compliance. For example, management might decide that certain benign resources will be substituted for toxic substances. Audits conducted by individuals already on staff are useful in determining whether these, and more complicated, policies are being followed.

Audit of Environmental Management Systems

Almost the entirety of the management structure of an environmentally intensive firm (such as chemical manufacturing), which means policies, practices, procedures, processes, and resources directed at environmental responsibility, should be adequate to ensure external and internal compliance. This is exactly the case with a well-implemented ISO 14001 EMS, and a full-scale audit will reveal whether the intent of the EMS is being accomplished.

Audit of Risks and Liabilities of Property Transfers

Audits of the environmental liabilities or impacts of potential acquisitions or divestitures may be conducted prior to these transactions. This is a highly valuable risk-management tool. Under the Superfund provisions, a party may be liable for the past offenses of a piece of property.

Specifically, according to Campbell, the Superfund legislation places liability on four classes of landowner and land user: (1) owners and operators of sites where hazardous materials have been dumped; (2) owners and operators of these sites *at the time of the disposal*; (3) those parties who generate hazardous materials; and (4) those who have transported or who transport hazardous materials to the site. "Thus," she writes, "current owners of land or facilities can be held responsible for environmental contamination regardless of whether or not they caused the contamination" (p. 51).

These are also called *transactional* audits.

Audit of Treatment, Storage, and Disposal of Hazardous Wastes

One of the major ways to make a negative environmental impact is to have anything at all to do with hazardous wastes. The Resource Conservation and Recovery Act (RCRA) is aimed at governing all activities related to hazardous waste, from production to disposal. The EPA requires that these activities be accounted for in writing.

Audit of Pollution-Prevention and Waste-Minimization Programs

Ideally, of course, waste should be minimized at its source. This would render legislation such as RCRA obsolete. The focus, Campbell writes, should be on prevention rather than on later cleanup. ISO 14001 subscribes to this tenet, thanks to input from the EPA.

Auditing these programs is a very proactive way to determine the efficacy of waste-minimization efforts at the source.

Audit of Financial Accounting for Environmental Liabilities

These are audits of how a company goes about paying for its environmental misdeeds or other obligations.

Audit of the Environmental Performance of Other Parties

We referred to this above as second-party auditing. The principle at work here is that products and services obtained by an organization should be subject to the same standards that the organization follows. This occurs if an organization or facility is conducting truly comprehensive environmental management.

While it is not always possible to arm-twist suppliers into conforming with particular environmental management principles, such as ISO 14001 or others, it is appropriate for a facility, according to the International Chamber of Commerce, to

> promote the adoption of these principles by contractors acting on behalf of the enterprise, encouraging and, where appropriate, requiring improvements in their practices to make them consistent with those of the enterprise; and to encourage the wider adoption of these principles by suppliers (p. 54).

It is also possible, Campbell writes, that small subcontractors and suppliers will be encouraged by the examples set by larger organizations to undertake these improvements by their own initiative. Principles of corporate risk management require consideration of subcontractors' environmental and other practices.

The CERES Principles

Just as environmental auditing didn't appear with the birth of ISO 14001, neither did the notion of environmental responsibility. Although larger companies have been reluctant (Campbell, p. 29) to adopt them, the so-called CERES—Coalition for Environmentally Responsible Economies—principles have found favor with a few major corporations, such as General Motors and Sun Company.

The principles were written by a confederation (p. 29) of U.S.-based environmental groups, investors, trustees of pension funds, and religious organizations. They were originally called the Valdez Principles after the 1989 oil spill. The reason these commonsense commitments haven't found universal corporate favor is that they are among the strictest environmental pledges on record. CERES requires companies to disclose their environmental performance records, restore the environment (an admittedly generic goal), minimize pollution, and conserve natural resources.

Whether more companies eventually adopt the CERES commitment, ISO 14001, or any other standard, there is no question that management of some companies has advanced from a reactive to a proactive environmental stance (p. 46).

Here are the 10 CERES principles (p. 30):

1. *Protection of the biosphere*

   We will reduce and make continual progress toward eliminating the release of any substance that may cause environmental damage to the air, water, or the earth or its inhabitants. We will safeguard all habitats affected by our operations and will protect open spaces and wilderness, while preserving biodiversity.

2. *Sustainable use of natural resources*

   We will make sustainable use of renewable natural resources, such as water, soils, and forests. We will conserve nonrenewable natural resources through efficient use and careful planning.

3. *Reduction and disposal of wastes*

   We will reduce and where possible eliminate waste through source reduction and recycling. All waste will be handled and disposed of through safe and responsible methods.

4. *Energy conservation*

   We will conserve energy and improve energy efficiency of our internal operations and of the goods and services we sell. We will make every effort to use environmentally safe and sustainable energy resources.

5. *Risk reduction*

   We will strive to minimize the environmental, health and safety risks to our employees and the communities in which we operate through safe technologies, facilities and operating procedures, and by being prepared for emergencies.

6. *Safe products and services*

   We will reduce and where possible eliminate the use, manufacture, or sale of products and services that cause environmental damage or health or safety hazards. We will inform our customers of the environmental impacts of our products or services and try to correct unsafe use.

7. *Environmental restoration*

   We will promptly and responsibly correct conditions we have caused that endanger health, safety, or the environment. To the extent feasible, we will redress injuries we have caused to persons or damage we have caused to the environment and will restore the environment.

8. *Informing the public*

   We will inform in a timely manner everyone who may be affected by conditions caused by our company that might endanger health, safety or the environment. We will regularly seek advice and counsel through dialogue with persons in communities near our facilities. We will not take any action against employees for reporting dangerous incidents or conditions to management or to appropriate authorities.

9. *Management commitment*

   We will implement these Principles and sustain a process that ensures that the Board of Directors and Chief Executive Officer are fully informed about pertinent environmental issues and are fully responsible for environmental policy. In selecting our Board of Directors, we will consider demonstrated environmental commitment as a factor.

10. *Audits and reports*

We will conduct an annual self-evaluation of our progress in implementing these Principles. We will support the timely creation of generally accepted environmental audit procedures. We will annually complete the CERES Report, which will be made available to the public.

More on the Role of the Internal Auditor

Internal auditing is not limited to environmental auditing. It is actually a "broad-ranging appraisal function" established within many organizations. Campbell quotes Sawyer (1981),* who defines internal auditing as "an independent appraisal of the diverse operations and controls within an organization to determine whether acceptable policies and procedures are followed, established standards are met, resources are used efficiently and economically, and the organization's objectives are being achieved" (p. 8).

A variety of professionals are used for internal auditing, as it is a highly specialized function. How specialized is discussed in the Institute of Internal Auditors' (IIA) *Standards for the Professional Practice of Internal Auditing*. Section 300 (p. 8) of this document establishes the internal auditor's precise responsibilities. Section 320 says the following, according to Campbell: "Internal auditors should review the systems established to ensure compliance with those policies, plans, procedures, laws, and regulations which could have a significant impact on operations and reports, and should determine whether the organization is in compliance."

As will become apparent, lack of compliance with applicable laws, regulations, and procedures can have a significant impact on operations and reports in both the long and the short run. The risks of regulatory noncompliance, of course, are the most serious and can include jail time. But noncompliance is not the worst offense—that distinction belongs to short-term environmental degradation and human suffering, according to Campbell. Internal auditing *professionals* are needed to head off the risks inherent in these types of corporate and organizational behavior.

The IIA's Scope of Work general standard establishes the internal auditor's responsibility for reviewing an organization's system of internal control (p. 60): "The scope of the internal audit should encompass

---

*Sawyer, Lawrence, *The Practice of Modern Internal Auditing*, 1981.

the examination and evaluation of the adequacy and effectiveness of the organization's system of internal control and the quality of performance in carrying out assigned responsibilities."

Internal control should have five primary objectives, each of which can be applied to the environmental realm, according to Campbell (p. 60):

- The reliability and integrity of information
- Compliance with policies, plans, procedures, laws, and regulations
- The safeguarding of assets
- The economical and efficient use of resources
- The accomplishment of established objectives and goals for operations or programs

While there is little empirical evidence concerning internal auditing's impact upon environmental auditing in general, some useful information on companies' specific practices is available.

A 1992 Price Waterhouse survey of 236 firms, selected for their significant environmental liabilities, indicated the following levels of internal auditor use for environmental audits (p. 66):

- Audits of financial accounting for environmental sites: 33 percent
- Audits of compliance with environmental laws and reporting requirements: 40 percent
- Audits of compliance with company environmental policies and procedures: 58 percent

The survey revealed that the environmental audit function is most often "owned" by a highly technical and specialized unit within the company, such as environmental affairs, environmental engineering, safety, health, or environmental assessment. When such a particular segment of the organization takes control of internal auditing, it can make these contributions (p. 67):

- Monitoring the environmental auditing process
- Bringing audit knowledge, expertise, and an exchange of ideas to the process
- Contributing a risk model to set priorities for audits
- Training the technical environmental audit team in areas such as statistical sampling and audit planning
- Providing staff to the environmental audit team

- Paying attention to specific environmental issues during a general audit

Toward this end, intentionally or otherwise, "the IIA has taken several steps to increase the internal auditing profession's awareness of environmental issues and environmental auditing"—for example, by offering courses on environmental auditing (p. 68).

Campbell's own research, which included 186 out of 599 companies contacted, indicates the following reasons for environmental audits (p. 140): compliance with environmental regulations, 92.9 percent; compliance with company environmental policies, 79.2 percent; environmental management system audits, 65 percent, environmental liabilities and risks of property transfers, 88 percent; programs and facilities for the treatment, storage, and disposal of hazardous waste, 89.1 percent, pollution-prevention and waste-minimization programs, 81.4 percent; and financial accounting for environmental liabilities, 66.1 percent.

Of these companies, 80 percent conduct "second-party" audits of other parties; the likely impetus for this is potential property acquisition and a desire to avoid Superfund penalties for past sins (p. 141).

While 90 percent of the companies Campbell surveyed conduct *environmental* audits, more than 40 percent reported no internal audit involvement in these reviews. She calls this her most striking finding (p. 151). The 90 percent figure can perhaps be attributed to the fact that corporate officials reported that environmental protection was a much higher priority in 1994, during Campbell's research, than in 1991. On a scale of 1 to 7, with 7 being the highest priority ranking, environmental protection rated a mean score of 6.26, up from 5.36 in 1991.

Conclusion

The benefits of environmental auditing are not only unquestioned but lauded by many scholars and industry leaders (p. 55). The most prominent benefit is the assured compliance with applicable regulations and legal obligations.

Conversely, an organization may also benefit by avoiding fines and other penalties resulting from *noncompliance* with these requirements. Tangentially, financial reporting requirements, such as those laid out by the SEC, are more easily adhered to when environmental activities are closely monitored.

Of nearly equal importance is the preventive or proactive stance a company displays by actively focusing on prevention and minimization at the source of a product or service, leading to possible long-term cost savings.

But such open concern for the environment may have its risks, especially for firms which have significant and unchanged environmental impacts. "Fear of legal exposure has discouraged some companies from implementing environmental management systems that document environmental problems," Campbell writes (p. 57). "Organizations are concerned that their environmental auditing reports may be used against them in court." The EPA's promised case-by-case method of determining who must turn over audit reports has not put all corporate minds at ease.

One reason some corporations have preferred to channel environmental documentation through their legal departments has been the belief that such papers would be protected by attorney/client privilege, leading to the notion that environmental compliance is merely a legal aspect of operation best left to the lawyers. But state and federal court rulings have lifted this protection (p. 58).

Another understandable reason some firms may be reluctant to conduct environmental auditing is their fear of being financially unable to resolve significant environmental impacts.

Although regulatory compliance is the chief reason for conducting environmental audits, it is not the only prominent one. Campbell writes that about 80 percent of the firms she surveyed focus on compliance with company policies, pollution prevention, and waste minimization. Whatever the reason for the audit, Campbell's research clearly indicated that for most, if not all, companies, its benefits far outweighed its risks. On the other hand, internal auditors had only a limited role in environmental auditing among her respondent facilities—a fact that doesn't necessarily point to lax environmental management, for "there is no universal best organizational form for an environmental auditing program" (p. 216).

Here, again according to Nestel and colleagues, is a list of tips for adequate reporting:

- Develop a communications strategy, considering the use of annual environmental reports, customer knowledge of your performance, and other actions.
- Begin evaluating your company's environmental management process and tracking environmental spending trends.
- Consider benchmarking your company's environmental management process/practices against those of competitors and/or of companies widely viewed as following "best management practices."

- Look to your suppliers and customers as well as to your competitors for important trends, and adopt a global perspective on development.
- Think about ways to reengineer your management process so that environmental considerations are integrated directly into the business management/product development processes. Look for performance development tools that enhance your company's effectiveness and productivity in managing environmental matters.
- Focus on ways to competitively differentiate your company by adopting cost-saving approaches like waste minimization/pollution prevention, as well as by evaluating new revenue-generating market opportunities.

"The key issue," the article states, "is not whether your company will need to increase its environmental reporting, but how your communications and reporting strategy can be developed and implemented to enhance your company's image and return-on-investment."

3
Don't Feel Like You're Out of Your "Element"

The reason for the numbering system in this chapter is that all of the *mandatory* ISO 14001 elements appear in clause 4 of the ISO 14000 series (clauses 1 to 3 cover scope, normative references—there are none as of this writing—and working definitions of critical terms). ISO 14001 is actually the size of a short booklet and contains three "annexes," documents which help clarify and provide guidelines to the ISO 14001 specification.

In this section, we'll use an italic typeface for the wording of the elements, followed by our discussion of the application of each. We should also say at the outset that, because companies vary so widely in size, complexity, and function, it is impossible to adapt this discussion to every type of organization that has an interest in implementing an EMS or registering to the ISO 14001 standard.

What we have attempted here is as straightforward an interpretation of the elements as possible without any specific type of business in mind. Note that wherever it appears in the specification, the word *shall* indicates a requirement which *must* be met if the EMS is to qualify for ISO 14001 registration.

Element 4.1 General

The organization shall establish and maintain an environmental management system, the requirements of which are described in this clause.

The two buzzwords here are *establish* and *maintain*.

The nature of your environmental management system, hereafter referred to as the EMS, must first be *established,* meaning that you must decide what it will look like and how it will operate.

Only after making these decisions can you implement, or put into place, the EMS. Following effective implementation, the EMS must be *maintained*—a process which must embrace continual improvement.

Element 4.2
Environmental Policy

Top management shall define the organization's environmental policy and ensure that it:

a) Is appropriate to the nature, scale, and environmental impacts of the organization's activities, products and services. The policy you develop must take into consideration the size and complexity of your operation and your company's effect on the environment. Going back to our Exxon example, if you ship crude oil all over the world, your policy must address how you will conduct that activity and simultaneously reduce any negative impact it might have on the environment. A one-sentence policy that says, "We'll protect the environment" is an unacceptable oversimplification.

b) Includes a commitment to continual improvement and prevention of pollution. You must indicate that not only will you reduce the environmental impacts you're already making, you'll continue to strive to improve your activities, products, and services until you've minimized your environmental impacts.

c) Includes a commitment to comply with relevant environmental legislation and regulations, and with other requirements to which the organization subscribes. This item should be self-explanatory.

d) Provides a framework for setting and reviewing environmental objectives and targets. In this section of your policy, you are asserting that you have the basis for identifying and executing improvement mechanisms. Before you can do this, you must know what needs to be improved.

In other words, your environmental policy cannot be an arbitrary set of goals. It must address those activities that have the potential for a negative impact on the environment and make a commitment to ameliorate each. Your environmental policy need not be long or technical. It must demonstrate, however, an awareness of what your EMS must accomplish.

e) Is documented, implemented, maintained and communicated to all employees. This part of your policy indicates a commitment to train employees so that they are aware of their role in fulfilling your organization's environmental objectives.

f) Is available to the public. Hang your environmental policy in the lobby where it is available for public perusal, or included printed copies with vendor's contracts or on retail products. These are easy ways to make the policy available to the public.

Element 4.3 Planning

Element 4.3.1 Environmental Aspects

The organization shall establish and maintain a procedure to identify environmental aspects of activities, products and services that it can control and over which it can be expected to have an influence, in order to determine those which have or can have significant impacts on the environment.

If your organization is doing something that affects the environment, and it's something you can control, you should know about it.

An environmental aspect is something that has the potential for interacting with the environment. If it does interact, this is an environmental impact, which may be negative or positive.

It becomes a *significant environmental impact* if the wider community and legislative bodies have demanded that something be done about it.

In terms of conforming to this element, then, it helps to keep up with the activities of environmental public interest groups, stay abreast of legislation, and be aware of whether your activities are affecting the outside world. Put someone in charge of monitoring these matters.

The organization shall ensure that the aspects related to these significant impacts are considered in setting its environmental objectives. The organization shall keep this information up-to-date.

The discovery that your company is having a significant environmental impact shouldn't occur in a vacuum. If you now know, for example, that scraps from the manufacture of your bicycle tires are polluting the area behind your plant, you should use this information to set goals for the future prevention of pollution.

Resolve to periodically check for these and other environmental aspects.

Elements 4.3.2 Legal and Other Requirements

The organization shall establish and maintain a procedure to identify and have access to legal and other requirements to which the organization subscribes (that are) directly applicable to the environmental aspects of its activities, products or services.

Staying abreast of complicated and constantly evolving environmental legislation can be fairly overwhelming. Even if your environmental background is strong, you (or the individual to whom you give this responsibility) will face a daunting task.

Most legislation is extremely abstruse. A law that sets forth provisions for something as simple-sounding as handgun possession may be 200 pages long and filled with convoluted and seemingly irrelevant language. Imagine what a piece of legislation that sets pollution limits would look like.

Also, because the law is an ever-changing and growing entity, what is applicable this year may not be applicable next year or three years hence. What actually changes may be a small fraction of the existing law or the whole thing. And these changes, unless they're earth-shattering, are not likely to make the front page.

It is also helpful, in the interest of being proactive, to keep track of emerging issues and trends which will affect you. While this need not necessarily be a full-time job, it is one that must be approached thoughtfully and diligently.

So how should you start?

While this is not required by ISO 14000, we agree with the provision of BS 7750 (ISO 14000's forerunner) that says a facility should keep a *register* of legislative, regulatory, and other policy requirements. Such a register can be a simple chart that lists the name of a law and its specific applicable requirement, the area of a facility that is affected and what the effect is, and what current facility procedure must be altered in order to ensure compliance.

Don't Feel Like You're Out of Your "Element"

David Hunt notes that it is not sufficient simply to acknowledge in such a register that a facility is governed by Law X—that facility must understand how Law X is applicable:*

> BS 7750 requires not a register of legislation itself, but [an understanding of] *requirements* [author's original emphasis]; it is the practical implications in terms of day-to-day activities which need to be demonstrably understood. The purpose of the EMS is to facilitate compliance with these requirements, and it is important to keep this end in sight (p. 162).

An example of what a register might look like appears in Figure 3-1.

With our society's rapidly advancing computer and other technology has come a proliferation of ways to keep track of emerging legislation, regulations, and other requirements. Some publications are produced with the sole purpose of tracking such legislation; others are targeted to a specific readership, such as professors or business leaders.

It is unlikely that any given organization needs to review all the available information. Identifying which materials are most relevant to a facility is a good way of whittling down the volume of "required reading" (p. 163).

Kleen-Air Register of policy, legislative, and regulatory requirements			
Item #	Name of requirement	Area(s) affected	Relevant Procedure
1.1	Water Regulatory Act	Wastewater discharge	KA-1-5a
1.2	Environmental Act	Recycling of raw materials	KA-71-2-3
1.3	Worker Safety Act	Maintenance of uniforms	KA-441-B
1.4	Policy F10 on filter types	Purchasing	KA-36

Figure 3-1. One good way to keep track of relevant legislative, regulatory, and policy requirements is by way of a chart called a *register*.

---

*David Hunt, *Environmental Management Systems: Principles and Practice*, McGraw-Hill, New York, 1995.

Just as it is unnecessary to read every law journal, article, etc., it is unnecessary to ask every employee to read the material you do choose to use. The management representative (MR), who is ultimately responsible for seeing that the ISO 14000 Environmental Management System is implemented, should spearhead the collection of relevant literature and digest as much of it as possible. He or she may also choose to select one or more additional persons, from the company's library or legal department, to read the rest.

Others who may be rightfully assigned this responsibility include personnel from business development, external communications, waste handling and disposal, and research and development (p. 163).

You as an executive may even choose to do some "preliminary sifting" (p. 163) of the literature you procure before passing it along to your MR. Some of the material may strike you as inappropriate, too vague, etc. Whatever method of distribution you choose should be relatively simple and include a distribution list that ensures that all appropriate information is gleaned.

Naturally, not all sources of information are subject to this type of strict organization. Information learned at meetings and conferences and informal conversations are represented on paper for all to see. That is not to say, however, that such seminars, meetings, and other business gatherings should be avoided or taken lightly. Instead, decide who best to send to these meetings and see that they record and distribute relevant data. This policy, promulgated by the MR, guarantees that no current environmental requirements will be overlooked (p. 163).

Don't treat the research, in whatever form you receive it, as something which may be skimmed once or twice by "relevant personnel" and then tossed into a pile in case anyone should be interested in looking at it later. Information about environmental requirements should be controlled like any other ISO 14000 EMS documentation.

Institute procedures for the storage and ready retrieval (p. 163) of this documentation. This can be done using computer software or other media, depending upon which the organization finds more useful. How you document your holdings and where you keep that documentation will depend, again, upon what works best for your facility. But a certain amount of paperwork, such as correspondence from regulatory personnel, may be unavoidable.

Should you choose to write an environmental management system manual, which we recommend, you may want to include a master list of all legislative and regulatory documentation that you are currently using or direct the reader to the documentation (p. 163). In any case, it should be easily accessed.

Earlier, we referred to the concept of keeping a register of applicable environmental requirements. The commitment to assemble or compile such a document should include documented procedures for updating the register in order to accommodate the ever-changing world of environmental legislation and regulation. We will soon discuss document control, one of the key aspects of the specification standard ISO 14001. Updating your register is one of the best ways to comply with the document control element; out-of-date or redundant requirements can sink your organization's efforts to maintain environmental compliance (p. 163).

As we shall see later, some industries may choose to abide by still other requirements that may have nothing to do with the government's command-and-control tactics. The chemical industry, for example, follows a voluntary model called Responsible Care.

The Catch-22 of keeping up with environmental or any other category of legislation is that documents may become obsolete the minute they roll off the presses (or even before they are printed). This will be your dilemma, especially if you decide to rely upon commercially available legal guides. Ironically, these may be your most accessible and plentiful source of useful information.

There are organizations dedicated to maintaining updated information and sending it to subscribers. The information they provide, often available in loose-leaf form, is more expensive (p. 163) but more reliable. Using such a service removes the burden of doing independent and time-consuming research. A service like this may even alert subscribers to pending legislation and relevant trends.

For the technically well-versed, there are computer-based systems which can be accessed through a telephone line and modem at times that are most convenient for the user. As with many so-called on-line services, it may be possible to call up information by way of a keyword search. This is likely to be the most expensive option available.

If you plan to be truly diligent in your research, you may wish, in addition to one or more of the options just discussed, to keep in touch with trade associations and their newsletters—most likely a relatively inexpensive choice. Keep your eyes open for conferences and seminars held by governmental and other bodies. These are likely to be advertised in association publications.

Here, with credit to Hunt, is a partial list of available commercial publications and services:

Books: *Yearbook of International Environmental Law; European Community (now European Union) Environment Legislation; Regulating the European Environment; NSCA Pollution Handbook; Tolley's Environmental Handbook: A Management Guide; Environmental Law; Corporate Environmental*

Responsibility: Law and Practice; The Environmental Protection Act of 1990; Waste Management Law: A Practical Handbook; Wisdom's Law of Watercourses; Water Law: A Practical Guide to the Water Act of 1989; and *The Law of the National Rivers Authority.*

Booklets: Law firms' occasional reviews, such as Clifford Chance: *Environmental Law Guide* and Denton Hall: *A Guide to Environmental Registers.*

Loose-leaf subscription services: Barbour's Health, Safety and Environmental Index; Croner's Environmental Management; Croner's Waste Management; Garner's Environmental Law; Encyclopedia of Environmental Health Law and Practice; Manual of Environmental Policy: The EC (now EU) and Britain; and European Environment Law for Industry.

Computer-accessible systems: Enflex Info (ERM) and Silver Platter.

Newsletters and journals: *Environment Business; The ENDS Report; Environment Information Bulletin; Environmental Law Monthly; European Environmental Law Review; Environmental Law Reports;* and *Environmental Law Brief.* (Some law firms' environmental law departments also publish newsletters. See Simmons and Simmons' *Environmental Law Newsletter* and Nabarro Nathanson's *Environmental Law Matters.*)

Element 4.3.3   Objectives and Targets

The organization shall establish and maintain documented environmental objectives and targets, at each relevant function and level within the organization.

When establishing and reviewing objectives, an organization shall consider legal and other requirements, its significant environmental aspects, its technological options, and its financial, operational and business requirements, and the views of interested parties.

The objectives shall be consistent with the environmental policy, including the commitment to the prevention of pollution.

Objectives and targets must be established for each level and function within the organization that is *related to environmental performance.* They must stem, as would logically be the case, from your environmental policy—where you first state your commitment to sound environmental management.

Element 4.3.4   Environmental Management Program

The organization shall establish and maintain a program or programs for achieving its objectives and targets. It shall include:

1. *Designation of responsibility for achieving objectives and targets at each relevant function and level of the organization.*

2. *The means and timeframe by which they are to be achieved. If a project relates to new developments and new or modified activities, products or services, program(s) shall be amended where relevant to ensure that environmental management applies to such projects.*

The Environmental Management Program (EMP) is not the same as the EMS. The EMP is a component of the EMS and is the tool used to execute environmental objectives and targets.

Because you are likely to have more than one objective or target, you may have to implement a separate program for achieving each item. Alternatively, one EMP may be capable of covering several objectives.

You must decide upon and set a time frame in which each of these items is to be resolved.

For example, suppose that you have decided that you will reduce the release of chemical X into the river that runs behind your plant by 15 percent. Achieving that target is one separate program, for which you must appoint a responsible person and set a time frame.

If you have also decided that you will eliminate the leakage of chemical Z into the drain on the shop floor of your bicycle-tire factory, a plan must be implemented for that as well.

Element 4.4
Implementation and Operation

Element 4.4.1 Structure and Responsibility

Roles, responsibilities and authority shall be defined, documented and communicated in order to facilitate effective environmental management.

It is imperative that you have a flowchart, organizational chart, or some other graphic depiction of who is responsible for which tasks within the context of your EMS.

Everyone should know who has been put in charge of these various activities, so that when someone asks that something be done, it is clear to everyone just who has been granted the authority to carry it out.

Management shall provide resources essential to the implementation and controlling of the environmental management system. Resources shall include human resources and specialized skills, technology and financial resources.

Don't just appoint the first person who asks for a certain responsibility to that post; make sure you are assigning these oversight functions to people who are qualified. Go through your résumé files and screen prospects based on the extent of their technical backgrounds.

Naturally, if you are a small shop, there may have to be some overlap or extensive training.

Management must be committed to allocating any and all resources needed to bring operation of the EMS to fruition.

Top management commitment is an absolute prerequisite to the successful establishment, implementation, and maintenance of an EMS. Without this commitment, your EMS has no chance. Top management personnel do not necessarily have to get involved in the day-to-day nitty-gritty of implementation, but they have to understand the tenets of the standard and be prepared to lead the charge, providing visible leadership and applying pressure when necessary.

The organization's top management shall appoint a specific management representative (MR) or representatives who, irrespective of other responsibilities, shall have defined roles, responsibility and authority for:

1. Ensuring that the environmental management system requirements are established, implemented and maintained in accordance with this standard.

2. Reporting on the performance of the environmental management system to top management for review and as a basis for improvement of the environmental management system.

The management representative, as EMS ringleader, must

- Have the authority to commit resources.
- Get things done without a lot of conflict.
- Understand the concepts of establishing, implementing, and maintaining an EMS within the specifications of ISO 14001.
- Be appointed by management and represent the interests of management.
- Serve as a bridge between management and the rest of the company.
- Be held accountable for EMS development, implementation, and maintenance.
- Have unconditional management support.
- Have authority second only to the CEO and have access to, and backing of, the CEO.

- Have a genuine and passionate commitment to pollution prevention and other tenets of sound environmental management.
- Have the respect of other management personnel and other employees throughout the organization.
- Possess a detailed knowledge of management methods, as well as measuring and monitoring techniques.
- Earn the respect of the company at large.
- Periodically report to management on the progress of his or her activities.

At the other end of the spectrum, however, the MR may be little more than a coordinator with no power. Appointment to the position of management representative is not necessarily a promotion, even though the continual-improvement function of sound environmental management makes the MR's job a never-ending one.

Element 4.4.2 Training Awareness and Competence

The organization shall identify training needs. It shall require that all personnel whose work may create a significant environmental impact upon the environment have received appropriate training.

It shall establish and maintain procedures to make its employees or members at each relevant function and level aware of:

1. The importance of conformance with the environmental policy and procedures and with the requirements of the environmental management system.

2. The significant environmental impacts, actual or potential, of their work activities and the environmental benefits of improved personal performance.

3. The roles and responsibilities in achieving conformance with the environmental policy and procedures and with EMS requirements of the environmental management system including emergency preparedness and response requirements.

4. The potential consequences of departure from specified operating procedures.

Personnel performing the tasks which can cause significant environmental impacts shall be competent on the basis of appropriate education, training and/or experience.

When the decision has been made to implement an EMS, all employees should be made aware of what such a system means to the organization so that they can contribute to its effectiveness.

Employees should understand the concept of aspects versus impacts, the existence of objectives and targets, and the importance of carrying out these goals throughout the organization.

You may have to do some shifting around, new hiring, or additional training in order to install the best-qualified people in those positions with the most impact.

In order to be in compliance with this element, you must identify those training needs. They may be very diverse—what these needs are is entirely dependent upon the size, nature, and scale of your organization.

Element 4.4.3   Communication

The organization shall establish and maintain procedures for

1. Internal communication between the various levels and functions of the organization.

2. Receiving, documenting and responding to relevant communication from external interested parties regarding the EMS and environmental aspects.

The organization shall consider processes for external communications on its significant environmental aspects and record its decision.

Do you have some kind of interoffice mail? Is there some system by which personnel at your company can communicate important information or instructions to one another? Establishing such a system is required by clause (1).

Clause (2) refers to establishing some way of keeping track of the input from the public. Such input should be taken seriously, organized, and, if applicable, answered promptly.

Element 4.4.4   Environmental Management System Documentation

The organization shall establish and maintain information, in paper or electronic form, to:

1. Describe the core elements of the environmental management system and their interaction.

2. *Provide direction to related documentation.*

This is another good reason to compose an EMS manual, although you could probably fulfill this element of the standard without composing a complete book. If, in a few pages, either on hard copy or computer file, you can list the elements of your company's EMS, along with other relevant documentation, such as procedures and records, and where they can be found, you have complied with this requirement.

Element 4.4.5 Document Control

The organization shall establish and maintain procedures for controlling all documents required by the standard to ensure that:

1. *They can be located.*

2. *They are periodically reviewed, revised as necessary and approved for adequacy by authorized personnel.*

3. *The current versions of relevant documents are available at all locations where operations essential to the effective functioning of the system are performed.*

4. *Obsolete documents are removed from all points of use or otherwise assured against unintended use.*

5. *Any obsolete documents retained for legal or knowledge preservation purposes are suitably identified.*

Documentation shall be legible, dated (with dates of revision), and readily identifiable, maintained in an orderly manner and retained for a specific period. Procedures and responsibilities shall be established and maintained concerning the creation and modification of the various types of documents.

Knowing where your documents are usually includes answers to the following:

- When organization changes occur, are documents revised to accommodate these changes?
- Are the old documents thrown away when they are no longer needed?
- Have you decided how you're going to organize documentation?
- Are there scheduled intervals at which documents are reviewed?
- Are documents readily available? Do workers have access to them?

Element 4.4.6 Operational Control

The organization shall identify those operations and activities associated with identified significant environmental aspects in line with its policy, objectives and targets.

We previously defined the terms *environmental aspect* and *environmental impact* as they relate to long-range planning. But it's important that you know what your organization's environmental aspects and impacts are for another reason.

You may have some activity which, under specific conditions, can have a significant environmental impact. Should these conditions occur, you should have controls in place to head off that impact.

For example, nuclear power plants under certain conditions are subject to accidents. Controls are kept in place continuously to keep those conditions from occurring.

The organization shall plan these activities, including maintenance, in order to ensure that they are carried out under specified conditions by

1. *Establishing and maintaining documented procedures to cover situations where their absence could lead to deviations from the environmental policy and the objectives and targets;*

What will you do if these activities are not carried out when they should be? Are your plans in the events of such a situation documented and readily accessible?

1. *Stipulating criteria in the operational procedures.*

2. *Establishing and maintaining procedures related to the identifiable significant environmental aspects of goods and services used by the organization and communicating those relevant procedures and requirements to suppliers and contractors.*

You should also decide what activities are necessary to head off significant impacts made by goods and services you buy from external sources and alert the vendors and contractors involved to the need for such action.

Element 4.4.7 Emergency Preparedness and Response

The organization shall establish and maintain procedures to identify the potential for, and response to, accidents and emergency situations, and for preventing and mitigating the environmental impacts that may be associated with them.

The organization shall review and revise, where necessary, its emergency preparedness and response procedures, particularly after the occurrence of accidents or emergency situations.

The organization shall also periodically test such procedures where practicable.

Here we have another self-explanatory element of the standard, which is especially necessary for companies whose activities have the potential for significant environmental impacts.

You must document these procedures.

Element 4.5 Checking and Corrective Action

Element 4.5.1 Monitoring and Measurement

The organization shall establish and maintain procedures to monitor and measure on a regular basis the key characteristics of its operations and activities that have a significant impact on the environment.

An important note: Most monitoring and measuring takes place at the EMS level and does not entail peering into discharge tanks or looking directly at any other function related to the environmental performance of an organization.

Monitoring equipment shall be calibrated and maintained and records of this process shall be retained according to the organization's procedures.

This is a step for the part of your organization whose activities directly relate to environmental performance.

Keep records of calibrations in a handy location. Make this an ongoing priority and, if necessary, put someone in charge of overseeing the process. This is up to the MR.

The records need not be involved or elaborate. If a sticker attached to equipment in question works, that's sufficient.

The organization shall establish and maintain a documented procedure for evaluating compliance with relevant environmental legislation and regulations. This monitoring must be part of the management review process.

Element 4.5.2 Nonconformance and Corrective and Preventive Action

The organization shall establish and maintain procedures for defining responsibility and authority for handling and investigating non-conformance, taking action to mitigate any impacts caused and for initiating and completing corrective and preventive action.

Any corrective or preventive action taken to eliminate the causes of actual and potential nonconformance shall be appropriate to the magnitude of the problems and commensurate with the environmental impact encountered.

The organization shall implement and record any changes in the documented procedures resulting from corrective and preventive action.

You should decide who will be responsible for recognizing and correcting nonconformities. Whatever action this person takes should be adequate but not excessive.

Corrective action may require an adjustment of documented procedures.

Element 4.5.3 Records

The organization shall establish and maintain procedures for the identification, maintenance and disposition of environmental records. These records shall include training records and the results of audits and reviews.

EMS records shall be legible, identifiable and traceable to the activity, product or service involved. Records shall be stored and maintained in such a way that they are retrievable and protected against damage, deterioration or loss. Their retention times shall be established and recorded.

Records shall be maintained, as appropriate to the system and to the organization, to demonstrate conformance to the requirements of this standard.

Let's make a distinction between records and documents.

A *document* is any piece of paper which explains something. *Records*, on the other hand, are pieces of paper that support a function and prove it was performed.

All records are documents. Not all documents, however, are records.

Whenever we talk about records, we're referring to documents that are the results of some investigation or are proof of a procedure having been carried out.

As you can see, records are perhaps among your most important documents and should be treated with special care.

Element 4.5.4 Environmental Management System Audit

The organization shall establish and maintain a program or programs and procedures to carry out periodic environmental management system audits to be carried out, in order to

1. *determine whether or not the environmental management system*
   a) *conforms to planned arrangements for environmental management including the requirements of this standard;*
   b) *has been properly implemented and maintained;*
2. *provide information on the results of the audit to management.*

The audit program, including any schedule, shall be based on the environmental importance of the activity concerned and the results of the previous audits. In order to be comprehensive, the audit procedures shall cover audit scope, frequency, and methodologies, as well as the responsibilities and requirements for conducting audits; and reporting results.

You must conduct thorough and periodic audits of your EMS, not only in preparation for the registration audit, but also to periodically ensure that your EMS is functioning as prescribed. The results of these internal audits must be reported to management.

You must have a procedure in place to determine when and where in your organization these audits will occur. They must be conducted by qualified auditors.

Element 4.6 Management Review

The organization's top management shall, at intervals it determines, review the environmental management system, to ensure its continuing suitability, adequacy and effectiveness. The management review process shall ensure that the necessary information is collected to allow management to carry out this evaluation. This review shall be documented.

The management review shall address the possible need for changes to policy, objectives, and other elements of the environmental management system in light of environmental management system audit reviews, changing circumstances and the commitment to continual improvement.

The reason for management reviews can be summed up by the last two words of the element: *continual improvement.*

Because there are as many ways to plan and carry out these reviews as there are companies, the exact method of scheduling and conducting these reviews is something you will have to determine. You'll certainly want to hear from your management representative, who should be your best source of information about the EMS.

4
Finding a Registrar

Credibility of Registrars

Choosing an ISO 14001 registrar and checking that registrar's credibility is an uncomplicated process because there are only a few accreditation bodies which give registrars permission to register companies to the standard.

And, as ISO 14000 becomes widely adopted, more registrars are likely to hang out a shingle, eager to serve your registration needs.

The credibility of a registrar boils down to just a few key points, all of which are easy to investigate:

- Is the registrar qualified to grant registration to ISO 14001?
- Which body has accredited this registrar? In order for a registrar to have validity, it must be accredited by a recognized accreditation body. These boards formally "license" registrars to perform quality and management system audits. They are the United Kingdom Accreditation Service (UKAS), RvA (Dutch Council for Accreditation), Registrar Accreditation Board (RAB) in the United States, and International Register of Certificated Auditors (IRCA) in Great Britain.
- Is the registrar providing a complete description of the registration process, or might there be additional charges or add-ons?
- Is the registrar financially stable?
- Does the registrar use qualified auditors?
- Finally, does the registrar use auditors qualified to audit within the Standard Industrial Classification (SIC) codes that cover your business?

Auditing Philosophy

Another issue worth investigating is a registrar's auditing philosophy.

"My experience with registrars is that they run the gamut from people who are more interested in content to others who get wrapped up in form," said Joe Koretzky, the EMS facilitator at Sharp, the electronics manufacturer, in Washington state. Koretzky said that a company must decide how substantive a registration audit it wants to undergo.

Some registrars may be more cursory in their performance, which may make life easy for a client company with an unfocused implementation team, but would have long-term consequences in the event of a second-party audit, and in terms of a company's ability to maintain registration and control its environmental impact.

"I've heard of a registrar who didn't focus much on whether the content was sufficient; they were more concerned about whether, for example, corrective action forms were filled out properly than about the content of the corrective action. I would think that a good auditor should have a good understanding of the process of a desired corrective action to see if there is a recurrence of a situation—to see that the root cause has been identified," Koretzky said.

Before deciding upon a registrar, Koretzky said, a company should ask itself this question: Are we more interested in receiving a certificate or in installing a good management system?

Obviously, all the people you talk to are going to say that they're more interested in a good management system, but in many instances it's not true," he said. "They have to make that decision themselves."

He said he has seen companies that he personally would not have registered after seeing their environmental management systems—which says as much about a registrar as it does about the client company.

"I really don't care if a form is filled out properly," Koretzky said.

On the other hand, he conceded, "the failure to fill out a form on a recurring basis may indicate that the system may not be functioning well. Do you have a program that requires corrective action every four months?"

Even in those cases, Koretzky said, very few registrars will be willing to undertake the level of intervention necessary to detect those subtleties. That's why, if you want your registration certificate to mean something, you must hire a registrar who will be honest and thorough.

"The audit is only as good as the registrar is willing to make it," he said.

Koretzky said he has also seen companies betray *themselves* during registration audits by faking different aspects of their EMSs (for example, manufacturing a management review meeting's minutes and papering over other requirements)—things a registrar cannot detect. This only wastes money and, again, defeats the purposes of implementing an EMS.

Koretzky said that the best way to determine whether a registrar is right for you is by word-of-mouth referrals.

"It's tricky," he said, but not impossible, given the networking opportunities and settings where you can interact with other groups or individuals in your industry. He suggests using the Chamber of Commerce to find these opportunities.

When you have narrowed the field to just a few registrars, arrange to conduct interviews with their personnel. Considering the long-term relationship you must necessarily have with a registrar, through preaudits, preassessments, registration, and surveillance audits, you have every right to ask tough questions.

Not every registrar, of course, will agree with this notion. Koretzky recalled that a particular registrar was "taken aback" when he wanted to interview its lead auditors.

"Nobody had ever done that before. But my feeling is that if I'm going to establish a relationship, I want to talk to them before we get committed," he said.

Changing a registrar at any point during the registration process, especially after receiving certification, is cumbersome. It could mean returning to square one and starting the entire cycle over.

"Unlike pension plans," Koretzky said, "registrations are not portable."

Also, while it would be comforting to know that there is a full-time contact whom you can call with questions at any time, it will seem onerous if you ask for this as a provision of your contract. You should feel free to call the registrar with pressing questions as should any customer buying a service—but be careful how you word those questions.

Rather than ask how to solve a particular EMS problem—a question registrars are forbidden to answer—you should ask whether the registrar thinks a *particular* solution you are already considering makes sense.

"I would never put the auditor in the position of consultant," Koretzky cautions. "I would never jeopardize the auditor's objectivity. It depends on how you couch the question."

Take advantage of the time you spend with your auditors. If there is some time left after a preaudit meeting or preassessment, you can tidy up lingering issues then. Koretzky advises, however, that you not "go to the well too often."

"We have developed a good enough relationship with our auditors. If you don't go to the auditors too often, they can be a valuable resource," he said.

And it goes without saying that if any major changes occur in your company that affect your EMS, such as the addition of a new division, resulting in a widened audit scope requiring more auditors or person-days, this information should be communicated to the registrar without delay.

"It's critical that you don't surprise the registrar," Koretzky said.

Steps to Take before Hiring a Registrar

The January 11, 1996, issue of *Machine Design* offers further advice on finding a registrar. The advice is aimed at companies seeking ISO 9000 registration. But the authors acknowledge that ISO 14000 is based on ISO 9000, making their advice valid for either standard.

- Learn as much about ISO 14000 before hiring outside help.
- Seek a registrar that meets your company's bureaucratic needs as well as the needs of your overseas customers, if possible. Try to find a registrar that concentrates on companies in the same industry as your company.
- Interview as many registrars as possible. Pick their brains for information about the standard. But don't make this your chief mode of research on the standard.
- Don't be afraid to negotiate with your registrar. Remember that these are for-profit businesses which, like your company, are competing for business.
- Don't hire a registrar until you are ready to start the registration process.

Attempt an in-house preassessment in order to determine readiness. Some questions you will be asked on your application with a registrar are the following:

- What is your desired time frame for registration?
- Describe your business and any applicable SIC codes.
- What is your company's scope of operation?
- What is the size of your facility, and how many employees are there?
- What is the status of your existing quality system?
- What is the state of your quality system documentation?

Based upon this information, the registrar will prepare a price quote and an estimate of the time it will take to register your company—provided, of course, that all the necessary elements are in place. Your relationship with a particular registrar is finalized when you sign a contract for its services.

Registrar Accreditation

One major accrediting body, the Registrar Accreditation Board (RAB), based in the United States, began the first American-based ISO 14001 registrar accreditation program in early 1996. It grew out of a joint accreditation program for ISO 9001 registrars operated by RAB and the American National Standards Institute (ANSI). The two bodies now operate a single, joint ISO 14001 accreditation program.

John Donaldson, ANSI vice president of conformity assessment, said he had been optimistic about the eventual formation of a joint program.

"Our participation in a joint accreditation program is in the international interest and the country's interest," Donaldson said.

While awaiting word on the joint program, RAB operated its own registrar accreditation pilot program. The five participating registrars were giving RAB a crash course in environmental management requirements. Donaldson said that ANSI never started its own pilot ISO 14001 accreditation program.

Paul Fortlage, RAB vice president of operations, said, "We don't claim to say we know how to do environmental accreditation, because we normally accredit for quality management systems. We're going through new territory in the environmental area, and we needed to get the people in the environmental community fully involved in the process.

"There was great pressure from the marketplace for there to be one program," Fortlage said. "Two separate, independent programs would not be tolerated by those seeking accreditation."

Besides, he said, reinventing the wheel would have been counterproductive.

Fortlage said that the advantage for an American registrar of earning accreditation from an American accrediting body is largely cultural, because an American body is more likely to be familiar with and sympathetic to an American registrar's needs and to U.S. environmental regulations and codes.

The Raad voor Accreditatie (RvA), also known as the Dutch Council for

Accreditation, and UKAS, the United Kingdom Accreditation Service, also grant accreditation to American registrars. Officials of these two agencies, and of RAB, said that they welcome telephone calls from companies seeking to check the legitimacy of a registrar. Each maintains a list of the registrars it has accredited.

Here are those telephone numbers:

RvA: 011-31-30-239-4500

UKAS: 011-441-71-233-7111

RAB: 1-800-248-1946

The suitability of a registrar may also be measured against ISO Guide 62, "General Requirements for Bodies Operating Assessment and Certification/Registration of Quality Systems," which sets formal requirements regarding such essentials as liability coverage, confidentiality, qualifications of auditors, and an appeal process. Under the provisions of the guide, registrars must also prove that they are financially stable and impartial.

Guide 62 was prepared by the Committee on Conformity Assessment (CASCO) of the International Organization for Standardization and the International Electrotechnical Commission (IEC). It replaces Guide 48, titled "Guidelines for Third-Party Assessment and Registration of a Supplier's Quality System."

Registrars can also police themselves—if they so choose.

Casper van Erp of the RvA suggests that registrars appoint governing boards, consisting of interested parties representing a variety of professions, to give binding input into the certification process. He said such a board would serve in a capacity similar to that of a private corporation's board of directors.

van Erp said that participation should be solicited from industry representatives and from consumer and other public interest groups.

"A lot of companies," he added, "have found it impossible to employ the boards full time" and have found that doing so "wasn't suitable." Rather, the boards could convene "as needed" to oversee the formulation of policies critical to the certification process.

Before signing a contract, you and your registrar should thoroughly review the terms of the agreement. There are certain elements about which you should ask, such as under what conditions your contract can be altered in the event your needs change.

If you are uncomfortable, or your questions are not being answered satisfactorily, look elsewhere.

On the Horizon: Registrar Membership in Single-Auditor Certification Programs

Environmental auditors will have worldwide recognition when auditor certification bodies and registrars become members of the IATCA (the International Auditor and Training Certification Association) mutual recognition agreement.

IATCA has been working for years to devise uniform worldwide criteria for the training of auditors and for the certification or registration of auditors. Most major countries, including the United States, Canada, the United Kingdom, France, Japan, China, and Korea, have bought into the concept.

The only standard series used by the program at this writing (October 1996) against which to determine auditor competence is ISO 9000. But inclusion of ISO 14000 was scheduled for late 1997.

IATCA membership, which will benefit auditors (although only registrars may become members), carries the most stringent requirements—more so than any other certification or registration scheme. But membership carries an undeniable benefit.

"Until recently, if an auditor wished to work in the United States and the United Kingdom, he needed to be certified by the RAB in the United States and by IRCA in Great Britain. This puts the auditor to considerable expense and inconvenience," said IATCA secretary Noel Matthews. "What IATCA is saying is that if you're an American certified through RAB, or a Brit certified through IRCA, once those bodies become full members of the IATCA Mutual Recognition Agreement [which was expected to occur before the end of 1996], that certification will be recognized in virtually all the major countries of the world—the UK, Australia, Japan, China, Korea, Argentina, and so on.

"A company needn't satisfy itself that this person is competent," he said.

Large certification bodies or registrars like RAB and IRCA will continue to issue their own accreditations as well as certificates with the IATCA imprimatur to those auditors who meet IATCA mandates. It is hoped that all accreditations and registrations eventually will be based on the IATCA system.

"RAB has several thousand auditors on its books," Matthews said. "They have a moral responsibility to those people to continue their own program. A lot of those people will never be able to become IATCA auditors."

Certification bodies and registrars in smaller countries are expected to abandon their own programs and issue only IATCA-based certificates. The entire nations of Australia, New Zealand, and France are expected to use only IATCA standards by the year 2000.

The Joint Accreditation System of Australia and New Zealand, known as JAS-ANZ, will adopt the program immediately.

Although the objective is the same (certifying the competence of auditors), IATCA's certification requirements are the most stringent in the industry, which is why most auditors will never attain IATCA certification. It would be too disruptive to their careers to go back to square one for additional training, Matthews said.

Rather than the "bits and pieces" auditing currently required for most auditor certification, he said, which often does not require professional qualifications, candidates for IATCA certification must perform three full, acceptable audits under the supervision of a highly experienced examiner.

"That's never been done before by any certification body," Matthews said. "Most basically require that the auditor carry out audits, without any specific requirement of demonstrated competence.

"Not only does IATCA lay down the most stringent requirements for auditors, it requires auditors who seek IATCA certificates to carry out the audit function witnessed by a verifying auditor who's a very experienced, reputable person selected by a designated body as an examiner," he added. "The auditor actually has to carry out three audits under the supervision of that person and to that person's satisfaction."

Total auditing experience must total at least two years. In addition, applicants must have four years of direct, hands-on, increasingly responsible technical or engineering work experience ("real work, not just playing around or working in an office," Matthews said), or six years if they don't have a university degree. They must pass IATCA training and work an additional two years as an auditor.

"So it's a minimum of six to eight years, and it's more likely to be ten, so the world is not likely to become flooded with young (and inexperienced) auditors."

Because of the long and rigorous process of developing the rules and criteria for auditor certification, the program was still not in full swing as of 1996, although its development had been underway for years. Membership is granted—but only after a rigorous peer review—to certification bodies and registrars, who then issue IATCA certification to their auditors who have fulfilled the necessary requirements.

"So the credibility of their certificates means something," Matthews said.

Matthews said that the IATCA certification is needed to give the title of "auditor" more credibility than it commonly carries.

"The problem," he said, "is that a lot of auditors aren't all that good. The problem for the customer is: How do you tell the good from the not-so-good?"

A large delegation was working on a certification scheme for ISO 14000, although the one-year projection to complete the environmental auditor scheme may be a bit optimistic.

Paul Fortlage, RAB vice president of operations, said that several hundred applications for the IATCA certification had come in as of October 1996. But no action could be taken until the peer review process was completed.

"There's a real demand out there, and there's a lot of pressure to get it going," Matthews said. "The industry and regulators want a scheme that allows them to be able to use certified auditors anywhere in the world and to be confident that those auditors are competent for the work they are doing. If a management or environmental system certificate is to have real credibility, then the auditors who are on the front line granting the certificate must have credibility and competence. All interested parties need to be satisfied that the auditor will do the work to the required standard.

"That is the whole reason and justification for the IATCA program," he said.

5
The Moment of Truth: ISO 14001 Registration

Following the registration audit, the registrar may either approve, disapprove, or grant conditional registration to your EMS.

The registrar's job is relatively simple, compared to your job as the implementor of the EMS. But if you have prepared well for the registration audit, the process should turn out in your favor.

At the time of this writing, the actual ISO 14001 registration process had yet to be finalized, although not much was likely to change. It is highly comparable to the process for ISO 9000 registration.

Registration (Also Called Certification)

Here are the three possible registration "verdicts":

1. *Approval.* You will receive notice of pending registration if your system shows all the markings of an implemented ISO 14001 EMS–even if a few minor nonconformities were found.
2. *Conditional, or Provisional, Approval.* This status is granted if you have included most ISO 14001 elements in your system, but you do not

have a fully implemented, 100 percent ISO 14001 EMS. You may also win conditional approval if a number of minor nonconformities were found, thus breaking the chain of good, sound environmental management.

If corrective action is taken and can be substantiated in writing, your auditor may proceed with a recommendation for approval.

Important note: A registrar does not make an explicit recommendation for conditional approval. Rather, a recommendation for approval is withheld until corrective action has been taken and sufficient objective evidence is available to verify it. In other words, you have the benefit of the doubt in ISO 14001 registration.

3. *Disapproval.* You may have sufficient documentation but an environmental management system whose implementation (1) is poor or nonexistent or (2) fails to address the elements of ISO 14001. These are two major reasons to be denied certification.

Upon certification, you receive a handsome certificate for your outer lobby, and you are listed in a directory of ISO 14001–certified companies.

But What Does It Cost?

This is definitely a fair question. Again, because at this writing ISO 14001 registration was just getting underway, we can only use other standards as a guide.

When you enter the market for a registrar, you will find that there is a wide range of prices for registration services, depending on several factors. Each company and plant has its own unique characteristics, and these come into play in estimating costs.

One key piece of information, common to *all* audit schemes, is that you will pay for audit "person-days." You pay a predetermined amount for each day that *each auditor* spends at your facility.

Let's suppose that the size of your company requires a 16-person-day registration audit. If a registrar charges $1,500 per person-day, and you have four auditors for four days, multiply $1,500 by 4 and then multiply *that number* by 4, giving you a total cost of $24,000.

These prices are consistent no matter how a company chooses to fulfill the audit person-day requirement. In the case of that 16-person-day audit, a registrar may choose to send 16 people for 1 day, 1 person for 16 days, 2 people for 8 days, 8 people for 2 days, etc. You will pay the same amount.

# of Employees	Audit * Man-Days	Days On-site	Surveillance Audit Man-Days
1-4	2	1	1
5-9	3	2	1
10-14	3	2	1
15-19	3.5	2	1
20-29	4	2	1.5
30-59	6	4	2
60-99	7	5	2
100-249	8	6	2.5
250-499	10	8	3
500-999	12	10	4
1000-1999	15	12	5
2000-3999	18	15	6
4000-7999	21	17	7
8000+	24	20	8

* Including document review

Figure 5-1. The creators of ISO 14001 have not devised an audit "person-day" scheme that specifies how long a registrar may take to audit a facility seeking registration to that standard. For the time being, this audit person-day scheme used for ISO 9000 audits may also be used for ISO 14001 audits.

Since there is no established schedule of ISO 14001 audit person-days (as of August 1996), we use the ISO 9001 timetable, Figure 5-1 as a guide.

Some registrars will also offer you, at the same rate per person-day, the option of a preassessment—a "dress rehearsal" audit that reveals remaining weaknesses which can be corrected before the registration audit. Although it is not a mandatory step, we highly recommend it, as it helps assure that you will be as ready as possible for the actual event.

Keep in mind the possibility of travel expenses. Travel costs are usually added over and above the fee for registration, so you'll want to find out whether the registrar intends to fly auditors in from out of town or employs auditors in your area. The bottom line is that you should ask a registrar to give you an all-inclusive quote. Be thorough and demand a full accounting up-front. Beware of confusing added-on costs for office preparation and other services.

Finally, you will be charged a registration fee for the privilege of receiving your company's certificates of registration. The size of this fee depends upon the size of your company. It can range from $250 to $500.

Weigh the expense, however, against the potential savings in conserved energy, fuel, and water, waste reduction; and other aspects of your organization's operations. While the cost of registering to this or any other standard may give some companies pause, the long-term savings can make it all worthwhile.

6

SGS-Thomson: Blazing the Trail

The EMS was developed over a period of several years. As new regulatory requirements were enacted, compliance actions were identified and these actions were integrated into existing systems as much as possible. There were no negatives in implementing the EMS required for ISO 14001 certification.   JOE HESS
              SGS-Thomson Microelectronics

We are confident SGS-Thomson will be among the first in the country to become registered to ISO 14001 once the standard is formally published and this will happen automatically.   CAROL BROWN
              SGS-Thomson Microelectronics

Looking directly at the real-life saga of a company shooting for registration to the ISO 14001 specification may be instructive. Fortunately for us, a large company with offices scattered all over the world took this daring leap.

On January 3, 1996, the SGS-Thomson Microelectronics, Inc., semiconductor manufacturing facility in Rancho Bernardo, California, near San Diego, became the first U.S. plant to achieve registration to the ISO

14001 Standard for environmental management systems (EMS). Shortly before, on December 15, 1995, the same plant had also become the first in the United States to be verified to the European Union's Eco-Management and Audit Scheme (EMAS).

Environmental engineer Joe Hess said that other company facilities may pursue ISO 14001 registration. But, he added, EMAS's stricter validation (registration) requirements, including third-party audits that are comparable to an accountant's delving into the most obscure data within a business's financial records, make registration to EMAS a far more impressive imprimatur than ISO 14001 registration. This, in turn, means that registration to ISO 14001 is an additional expense for a less prestigious standard.

By its actions, SGS-Thomson has set the pattern for achieving ISO 14001 registration. It offers a model for other organizations to follow in properly implementing, managing, and documenting their EMS to reach compliance. Let's take a look at what this company did to achieve this major accomplishment.

The process that culminated in ISO 14001 registration began in late 1992 with a decision by corporate management to address environmental issues on a companywide basis—facility by facility. SGS-Thomson is a global independent semiconductor supplier whose corporate headquarters is in Saint Genis, France. At 16 separate facilities—in the United States, Europe, Southeast Asia, and North Africa—it employs 25,000 people, including 300 at Rancho Bernardo, which has been in operation since 1980.

The company designs, develops, manufactures, and markets a broad range of semiconductor integrated circuits used in a wide variety of microelectronics applications, including telecommunications systems, computer systems, consumer products, automotive products, and industrial automation and control systems. Its 1,500 customers include Alcatel, Bosch, Ford, Hewlett-Packard, IBM, Sony, Motorola, and Northern Telecom.

Throughout 1993, SGS-Thomson took a number of steps to deal with environmental matters systematically. These included the creation of corporate environmental strategies and international quality programs (ES-IQP), selection of an environmental consultant, appointment of a site environmental champion at each facility, a decision to implement EMS at all sites, a corporate mandate to achieve EMAS validation at all sites, the first draft of a company environmental policy, and environmental training for internal auditors, followed by environmental preaudits at all 16 sites.

The Rancho Bernardo site began preparing for certification to the ISO 14001 specification in 1994. According to site environmental champion Hess, this decision was made when it was apparent that the EMAS verifiers could also provide ISO 14001 registration. The entire registration and implementation process for both ISO 14001 and EMAS cost approximately $100,000.

Hess said he didn't know whether SGS-Thomson was planning to reap a return on that investment.

Patrick Hoy, the site's environmental manager, was Rancho Bernardo implementation champion, responsible for preparing the facility for EMAS and ISO 14001 certification. His duties included developing and implementing the EMS and ensuring compliance with environmental laws and regulations.

"The EMS was developed over a period of several years," said Hess. "As new regulatory requirements were enacted, compliance actions were identified, and these actions were integrated into existing systems as much as possible. There were no negatives in implementing the EMS required for ISO 14001 certification."

According to Hess, the site's EMS is based on strategic planning, operational control, and checking and corrective action.

Strategic planning is provided by general and specific policy statements, which are followed by objectives. Where practical, these objectives are measurable. Much of this strategic planning coincides with SGS-Thomson's corporate planning.

The company relies upon various process and facility specifications to provide operational control of site activities that may affect the environment or have a regulatory mandate. These specifications are controlled documents.

A production specification may describe the procedure for the proper disposal of contaminated wipes or the handling of a particular chemical. Other specifications, such as the storm-water pollution prevention plan, are developed to meet regulatory requirements. Monitoring conducted to ensure compliance with permit limits is also included in the operational controls program, along with associated abatement equipment, if applicable.

Checking and corrective action rely upon various internal and external mechanisms. Internally, a corporate audit is conducted every 18 months. There is also a documented site preventive-maintenance and inspection program, as well as weekly informal facility walkthroughs. Nonconformities identified through this process are tracked by an internal system. In addition, an annual audit is conducted using internal and external resources. This audit focuses on checking the site's EMS for conformance to the standard, as well as to both site and corporate environmental policies and objectives.

"Another hurdle to leap was getting employee buy-in for the EMS implementation initiative. We initiated a good-natured competition among sites to promote commitment," said SGS-Thomson spokesperson Carol Brown.

"As much as possible, personal responsibilities for the EMS have been written into the specifications. It is the employees' responsibility to be aware of the specifications that relate to their job function, with this information provided during new-hire orientation. Many of the objectives established in the EMS also have monetary benefits. We also conducted general EMS-awareness training for all managers and supervisors after registration," Hess added.

While Rancho Bernardo was developing its EMS during 1994, SGS-Thomson at the corporate level prepared the first draft of its ecological vision statement, the 10-point environmental policy (also called the Environmental Decalogue), undertook both internal environmental benchmarking and external benchmarking for comparison with four key industries, and carried out environmental corporate audits at all 16 sites to prepare for EMAS validation.

These internal corporate audits were based on the draft EMAS documents. They were performed by an audit team consisting of an independent environmental consultant, who served as lead auditor; a member of the corporate strategy department, and two environmental managers from other sites. These audits were scored, and the sites prepared the necessary documentation. The sites with the most developed EMS received the highest scores.

Rancho Bernardo's corporate audit occurred in early 1995 and set the stage for documenting the EMS and preparing for certification. The two essential elements in getting ready for registration were identifying environmental impacts and preparing the environmental manual.

"The key to any EMS is the mechanism employed for identifying aspects and impacts associated with a company's processes, products, and services. We followed the failure mode and effects analysis (FMEA) corporate standard developed for assessing the significance of identified environmental effects," Brown said.

"We reviewed all regulations that applied to Rancho Bernardo, including environmental legislation, corporate requirements, and media-specific permitting conditions, such as those relating to water, air, hazardous waste, solid waste, soil and groundwater, energy, chemical management, external noise, raw materials, new product processes, product planning, and emergency response planning. We ranked each impact according to its significance and its rank on our corporate list of overall environmental effects."

The determination of direct and indirect significant effects was based on SGS-Thomson's Corporate Environmental Procedures on the Assessment of Environmental Effects. They include continuous releases which are 5 percent or more above legal limits or corporate standards, accidental releases with the potential for very serious impacts, whether the site ranks among the top five local users of natural resources, continuous releases which cause concern in the surrounding community, and continuous wastewater releases which are not treated or abated.

The only significant impact discovered during the assessment was the use of single-walled buried piping to transfer waste acid to the acid neutralization unit. It could potentially have contaminated the soil during a rupture or leak. This nonconformity was corrected by moving the buried piping aboveground and double-walling it, an action which took several months to complete.

The audit also discovered that noise emissions from a liquid nitrogen plant were well within local regulations, but exceeded the corporate mandate. This plant will be replaced in mid-1997.

Hess admitted that this audit did not follow standard ISO auditing procedures. "In the future, a separate internal audit will be conducted," he said.

ISO 14001 and EMAS registration also brought about the development of both corporate and site environmental policies.

The *SGS-Thomson Environmental Decalogue* (see pages 79–82), which sets forth the corporate vision on the environment, was released for internal use and feedback in January, and was publicly released in September as part of Environment Day at corporate headquarters. Its requirements include EMAS verification at all sites by the end of 1997. The company's environmental policy was released as an official procedure in June.

At Rancho Bernardo, a general environmental policy covering health, safety, security, and the environment had been in existence for several years. In order to meet EMAS and ISO 14001 certification, several specific policy statements were also developed. These statements cover site activities, raw materials, noise, product design, accidents, information and training, compliance, future processes, energy, pollution prevention, the production process, vendors, contingency plans, environmental performance, and corrective action.

"We support compliance by major suppliers and contractors with environmental standards such as EMAS or ISO 14001. We cannot mandate certification, but we do recognize that environmental performance of our suppliers relates directly to the bottom line," Brown said.

SGS-Thomson at the corporate level selected Bureau Veritas Quality International (BVQI) as registrar. According to Hess, while the company cooperated with the auditors by ensuring that the necessary facility personnel were available for interviews, the working relationship was not a close one.

"We had very little time to develop a relationship," he said. "We would recommend creating a mechanism to ensure that a good working relationship is developed with the auditors. This may be accomplished through a series of meetings, during which the corporate and facility vision is communicated, while the auditors say how the audit may proceed and what they need for it."

BVQI conducted an initial EMAS audit at Rancho Bernardo in one day with a two-member team. Following this visit, the Rancho Bernardo facility developed an environmental manual to describe and document the site's EMS and ensure certification to EMAS and ISO 14001. According to Hess, existing documentation was sufficient for preparing the manual.

This controlled document describes specifications used throughout the facility to control environmental effects. Its content consists of policy, objectives, targets, program, responsibility and authority, training, communication, environmental aspects, operational controls, record keeping, and audit. The manual is used primarily by the health, safety, security, and environmental (HSS&E) department in its operations.

The EMAS validation audit was conducted in December by a three-person team over a three-day period. Minor changes were made in the environmental manual before the ISO 14001 registration audit, which was accomplished by one member of the EMAS certification audit team in one day.

The site's ISO/DIS 14001 certificate expired when ISO 14001 became an official international standard in the fall of 1996. The facility's certificate was automatically updated to that of the final standard.

"Practically, there was no difference between registering under the draft and under the final guidelines. This draft was the final one, and no change was expected. We were confident that SGS-Thomson would be among the first in the country to become registered to ISO 14001 once the standard was formally published," Brown said.

"We are extremely proud that this site has achieved this recognition. SGS-Thomson has been a leader in the development of corporate policies for the environment. This is a clear realization of these efforts. We believe validation to EMAS and certification to the ISO 14001 standards is evidence of this site's commitment to pursue the highest environmental standards available and to influence suppliers and contractors to do the same," Brown added.

"These approvals are a great reward for the staff, who have actively supported our environmental policies. To be the first in the United

States certified to ISO 14001 as well as approved to EMAS is a major achievement," said Rancho Bernardo plant manager Jack Mendenhall.

"It shows our commitment to the corporation's environmental vision to be recognized by all our stakeholders as a leader in environmental care, by exceeding regulatory requirements in both degree and timing wherever possible. It also shows our commitment to the corporation's environmental mission to eliminate or minimize the impact of our processes and products on the environment, maximizing the use of recyclable or reusable materials and adopting, as much as possible, renewable sources of energy in striving for sustainable development," said Hess.

Hess feels that Rancho Bernardo's experience offers many valuable lessons for companies seeking ISO 14001 registration.

"Our advice to other companies is to develop an EMS manual documenting procedures, keep the documentation updated, and ensure that the manual becomes a controlled document," he said. "It is important to identify the environmental effects of the facility, the operations that contribute to those effects, and the procedures, monitoring, and controls that are in place to manage these or potential effects. The policy statement and site's objectives should reflect the specific requirements of EMAS or ISO 14001."

But Hess also cautions: "Different facilities may encounter different circumstances during their ISO 14001 registration process."

SGS Thomson Microelectronics Environmental Decalogue

In SGS-Thomson we believe firmly that it is mandatory for a TQM-driven corporation to be at the forefront of ecological commitment, not only for ethical and social reasons, but also for financial return, and the ability to attract the most responsible and performing people. Our " ecological vision" is to become a corporation that closely approaches environmental neutrality. To that end, we will meet all local ecological/environmental requirements of those communities in which we operate, but in addition will strive to:

1.0 Regulations

1.1 Meet the most stringent environmental regulations of any country in which we operate, at all of our locations worldwide.

1.2 Comply with all the ecological improvement targets at least one year in advance of official deadlines at all our locations.

2.0 Conservation

2.1 *Energy*—Reduce total energy consumed (by our manufacturing, buildings, etc.) per million dollars sold by at least 5% per year with 25% reduction by end 1999.

2.2 *Water*—Reduce water draw-down (per million dollars sold) from local sources (conduits, streams, aquifers) by $\geq$ 10% per year, through conservation.

2.3 *Trees*—Reduce total paper and paper products consumption by 10% per year.

3.0 Recycling

3.1 *Energy*—Utilize alternative energy sources (renewable/co-generation) to a renewable degree. (At least 3 pilot plants by end 1999.)

3.2 *Water*—For all manufacturing operations, reach a level of 50% recycled water by end 1997 and 90% by end 1999.

3.3 *Trees*—Reach a usage level of 90% recycled paper, where we must use paper, by end 1995 and maintain that level.

3.4 *Chemicals*—Recycle the most used chemicals, e.g., for sulfuric acid, recycle $\geq$ 30% by end 1997 and 80% by end 1999.

4.0 Pollution

4.1 *Air Emissions*—Phase out all Class I ODS by end 1996. Contribute where we can to the reduction of greenhouse and acid rain generating gases.

4.2 *Water Emissions*—Meet the standards of the most restrictive community in which we operate, at all sites, for wastewater discharge.

4.3 *Landfill*—Achieve 100% treatment of waste at level 1 to level 4, of "Ladder Concept" preferability, with a half-life improvement goal of $\leq$ 1 year.

4.4 *Noise*—Meet a "noise to neighbors" standard at any point on our property perimeter of $\leq$ 60 dB(A) for all sites, from end 1995.

5.0 Contamination

Handle, store, and dispose of all potential contaminants and hazardous substances at all sites, in a manner to meet or exceed the strictest environmental safety standards of any community in which we operate.

6.0 Waste

6.1 *Manufacturing*—Recycle 80% of manufacturing byproduct waste (metal, plastics, quartz, glassware, etc.) with a half-life for reduction goal of ≤ 1 year.

6.2 *Packing*—Move to $> 80\%$ (by weight) recyclable, reused, or biodegradable packing materials (cartons, tubes, reels, bags, trays, padding) with a half-life improvement goal of ≤ 1 year.

7.0 Products and Technologies

Accelerate our efforts to design products for decreased energy consumption, and for enablement of more energy-efficient applications, to reduce energy consumed during operation by a factor of ≥ 10 by the year 2000.

8.0 Proactivity

8.1 Proactively support local initiatives, such as "Clean-up the World," "Adopt A Highway," etc., at each site in which we operate, and encourage our employees to participate. Undertake to lead in establishing such initiatives where none exist.

8.2 Sponsor an annual "environment day" at each site in which we operate, involving the local community.

8.3 Encourage our people to lead/participate in environmental committees, symposia, "watch-dog" groups, etc.

8.4 Include an "environmental awareness" training course in the SGS-Thomson University curriculum and offer it to suppliers and customers.

9.0 Measurement

9.1 Develop measurements for, and means of measuring progress/achievement on, above points 1.0 through 7.0, using 1994 as a baseline where applicable, and publish annual results in the "environment report."

9.2 Develop detailed means and goals to realize these policies, and include them in Policy Deployment by the end of 1995.

9.3 Continue the existing Environmental Audit and Improvement program at all sites.

10.0 Validation

Validate to EMAS standard, or equivalent, 50% of sites by end 1996, and 100% by end 1997. (In the event the validating authority is not available, this schedule can be delayed, but only for this reason.)

7

Surveillance Audits

After you have found a suitable registrar, signed a satisfactory contract, implemented a sound EMS, proven it, and attained registration status, periodic surveillance audits—also called maintenance audits—will be required in order to confirm that you have continued to conform to the standard.

Frequency of Surveillance Audits

How frequently ISO 14001 audits should ideally be done had yet to be decided at this writing. A consultant with whom we spoke predicted six-month intervals.

The frequency of surveillance audits, however, varies with the terms of registration. Your registration may be valid indefinitely, pending successful surveillance audits. Some registrars grant three- or four-year registrations and conduct surveillance audits at regular intervals within that period.

Still another option is a visit when the registration expires. We use the word *visit* rather than *audit*, because when this type of registration expires, your registrar will conduct a "reassessment" (or an assessment that is not quite a thorough audit but is not as cursory as a surveillance audit).

The Focus of Surveillance Audits

One thing you can expect a surveillance audit to cover is proof that you have continued to institute continual improvement and corrective action. The operative word is *continual,* said consultant/trainer Ralph Grover.

"If you stop paying attention to your ISO 14001 system," Grover said, "your surveillance audit will be bad. ISO 14001 demands continual attention. It's very easy to lose sight of that, in the sense that you have other business needs, and executive management, who is ultimately responsible for ISO 14001, has many different issues it has to deal with on a daily, weekly, monthly, yearly basis.

"Management has to pay attention to ISO 14001," he said.

The two main issues here are making sure the EMS works properly, and constantly paying attention to the control of environmental aspects. This, first and foremost, is why you have an EMS. Unfortunately, Grover said, there is no single effective way to prepare for this part of a surveillance audit.

Some parts are easier to nail down.

The Role of Stakeholder Complaints

One place surveillance auditors will want to visit is your stakeholder complaint file.

Stakeholders are those individuals and interest groups who are affected by various aspects of your operation. Because we all have a stake in preserving a healthy and livable environment, stakeholders include the neighbor who complains about noise or an unpleasant odor emanating from your plant. If you have not attended to these complaints, or if you have attended to them, but not to the satisfaction of the complaintiff, and that dissatisfaction has been documented, watch out—you're setting yourself up for a poor surveillance audit.

For the sake of your business, train employees to follow your company's procedure to stop whatever they are doing and listen carefully whenever an individual or interest-group-representative/customer calls to register a complaint about something your organization is doing—especially if it relates to an environmental impact. If necessary, print and reproduce a form that can be used to log these complaints. Keep the forms near the phone or at your service desk. Make sure there is a spot on this form for investigation results, the recommended course of action and, if possible, a target date by which this action will have occurred.

Appoint someone to handle complaints, someone who will make sure that these actions are initiated and completed.

Do it for the good image of your company—and for the sake of your registration document it!

8
What, You Worry? The Legal Aspects of ISO 14001 Registration

Registration to ISO 14001 is strictly voluntary. But if you do become registered, you should be aware that there may be legal consequences should you let your guard down and drift out of comformance.

Engaging a lawyer is not necessary to the registration process, although there are companies which choose to go that route. Should you choose to hire one, you may end up receiving some helpful advice.

Implementing and registering to ISO 14001 publicizes the fact that there are certain things about your organization which may adversely affect the environment. How the public (and your competitors) will react to these revelations is anyone's guess.

"You're potentially announcing to the world that you have some environmental impacts from your operations," said Grand Rapids, Michigan, lawyer Andrew Kok. "Before this point, those may or may not have been recognized by industry or the public."

Kok said a company may be biting off more than it can chew in terms of the legislative and regulatory requirements it pledges to follow.

ISO 14001 requires that you develop a plan to ensure compliance with applicable laws and keep abreast of changing developments on that front (see Chap. 3).

Some environmental insiders, Kok said, have talked about compelling companies to also meet requirements proposed by trade groups and industry—which could turn into a full-time job in itself.

"You are potentially creating a pitfall for the future if you cannot meet those management practices," Kok said.

We don't mean to suggest that you avoid ISO 14001 registration, especially if you do business internationally. It will become standard practice in some places. We are suggesting that you understand in advance how legally bound you will be once you are registered.

"If you say, 'We're going to make sure the floor is always neat and clean,' then fail to do it, and one day somebody slips and falls, you've set a level of care that you're not meeting," Kok said.

Richard Webber, a London attorney, said that although ISO 14001 evolved from the British standard BS 7750, lawyers in Britain have no involvement with the registration process and don't tend to advise about it, either.

Kok said that he has advised clients on environmental management system implementation, but not on ISO 14001 itself—but the considerations are the same. In either case, you should be well organized.

"I think it's a good practice to have a systematic approach to your environmental compliance issues," he said.

Confidentiality may be an important consideration for your EMS, Kok added, especially regarding EMS records that track the effectiveness of the system.

Nobody outside your company needs to know, for example, that despite your best efforts, you were unable to reduce the air emissions of a certain chemical for the first three weeks of implementation. The public relations fallout could be severe.

Remember, an EMS requires documents you may not have written before. Some of these items may expose your company to risk if they land in the wrong hands. These documents may also expose you to legal liability.

The actual legal implications in these cases are dependent upon the size and complexity of your organization as well as the extent of your required environmental compliance. A plastic injection molding company, for example, may not have as many potential environmental liabilities as the manufacturer of a toxic chemical.

But both state and federal inspections are strict, Kok said, so you must know both your obligations and your limits. And even if no harm has been done, fines and penalties for being out of legal and regulatory compliance could be "substantial," he said.

While lawyers are not needed to monitor the minutiae of an EMS, Kok advises that it is wise to let a lawyer scan draft versions of certain documents to confirm that you have not inadvertently left any doors open as far as liability is concerned.

"Ideally," he said, "the advice you'd get from a lawyer is appropriate and can be kept somewhat straightforward and simple. Companies shouldn't go overboard in trying to draft a system they'll have to live with."

EMS courses by Perry Johnson, Inc. explore legal issues and options for handling these issues.

Environmental Regulation and the EPA's Role

It's hard to be against the concept of environmental regulation. Much of the tangled web of legislation currently on the books is at least intended to benefit society at large. The problem with regulatory and legal compliance is that there are costs attached. These costs, according to Norman L. Weiss,* result from required actions and the ways companies choose to deal with those requirements (p. 97).

What makes justifying these costs difficult, Weiss writes, is that they are hard to understand and explain given the complexity and illogic of "today's regulatory environment."

Federal, state, and local interference in business (and, for that matter, much of society) is often seen as the antithesis to the smooth functioning of society. Some might even argue that there is a deliberateness about it all. What other theory, after all, could possibly explain the "endless paperwork" and pages of environmental regulation that govern so much of our behavior (p. 98)? What's more, there is no overriding philosophy or guiding principle that governs the development of this regulation in the United States. Like all matters political, regulation and lawmaking—even that regarding the health of our world—is the result of compromise and give-and-take (p. 98).

Environmental laws and other regulatory tools, such as rules decided upon by boards and commissions, are further complicated by the variety of environmental conditions, systems, and processes that are present in business enterprises and other entities today. A clear understanding of environmental requirements can be elusive, because they are built upon a foundation of enforcement, rather than one of public understanding (p. 98). They are of a command-and-control nature, requiring

---

*Norman L. Weiss, in Betty J. Seldner and Joseph P. Cothrel (eds.), Understanding Federal, State and Local Environmental Regulation, *Environmental Decision-Making for Engineering and Business Managers*, McGraw-Hill, New York, 1994.

specific actions. *How* those actions are to be carried out is a subject that seems to get lost in the paperwork, while the civil, criminal, and administrative penalties for violating those measures can be stiff. The subject is further obfuscated when the courts attempt to interpret laws, rules, and other statutes. Difficulties in attaining and understanding applicable legislation are even more of a problem in countries that don't publish periodicals that track such matters.

Still, environmental regulation remains a priority because of the widely accepted correlation between environmental protection and the public health (Weiss, 98).

The pendulum, however, has swung back from where, according to Michael E. Kraft and Norman J. Vig,* it was in the 1970s and 1980s, "when rapidly rising public concern about threats to the environment and government's eagerness to respond to this new political force initiated the 'environmental decade'" (p. 3).

"During the 1970s," write Kraft and Vig, "the United States, along with other industrial nations, adopted dozens of major environmental and resource policies, created new institutions such as the U.S. Environmental Protection Agency (EPA) to manage environmental programs, and greatly increased spending for them."

These policies, they write, even withstood the conservatism of the Reagan administration, as "the American public resisted efforts to weaken or reverse environmental policy."

Despite the professed environmental concern of the Clinton presidency (he was lauded for choosing U.S. senator and environmental author Al Gore as his running mate), the authors write, in the 1990s environmental initiatives were bound to undergo closer scrutiny as part of the "thoughtful search for more effective and efficient approaches" (p. 3).

"Environmental organizations are not likely to receive unconditional support for their agenda(s), even within the Clinton administration." Instead, the book states, the onus will be on state and regional government units to innovate environmental policy (p. 4).

Figures 8-1 and 8-2, as originally rendered by Kraft and Vig, show the executive branch agencies (Fig. 8-1) and House and Senate committees (Fig. 8-2) which share responsibility for environmental decision making.

Whatever the origin or progression of legislation, some of which Kraft and Vig deem absolutely necessary (such as that involving toxic waste issues that cannot be solved privately), local and state governments

---

*Michael E. Kraft and Norman J. Vig, "Environmental Policy in the 1990s," *Congressional Quarterly*, 1994.

What, You Worry? The Legal Aspects of ISO 14001 Registration

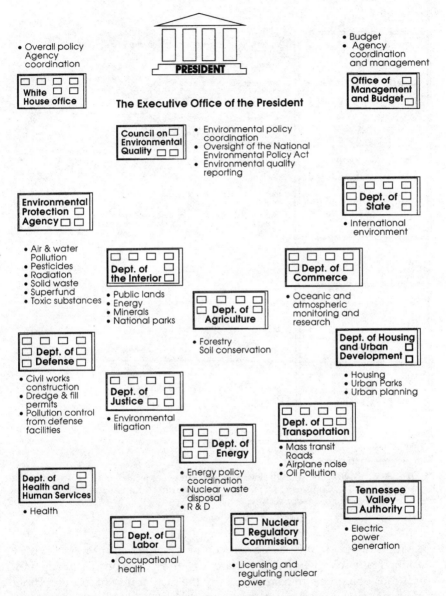

Figure 8-1. Several cabinet departments and other executive branch agencies of the U.S. federal government have responsibility for environmental policy.

HOUSE

Committee	Areas of Responsibility
Agriculture	agriculture in general, soil conservation, forestry, pesticide policy
Appropriations	appropriations for all programs
Energy and Commerce	Clean Air Act, nuclear waste policy, safe drinking water, Superfund, hazardous waste, and toxic substances.
Merchant Marine and Fisheries	National Environmental Policy act, oceanography and marine affairs, coastal zone management, fisheries and wildlife, wetlands
Natural Resources	public land, national parks and forests, wilderness, energy, surface mining, nuclear waste policy
Public Works and Transportation	water pollution, rivers and harbors, oil pollution, water power
Science, Space, and Technology	Environmental research and development, energy research, science and environmental issues

SENATE

Committee	Areas of Responsibility
Agriculture, Nutrition, and Forestry	agriculture in general, soil conservation, forestry, pesticide policy
Appropriations	appropriations for all programs
Commerce, Science and Transportation	coastal zone management; marine fisheries; oceans, weather, and atmospheric activities; technology research and development
Energy and Natural Resources	energy policy in general, nuclear waste policy, mining, national parks and recreation areas, wilderness, wild and scenic rivers
Environment and Public Works	air, water, and noise pollution; toxic and hazardous materials; Superfund; nuclear waste policy; fisheries and wildlife; ocean dumping, solid waste disposal; environmental policy and research in general

Figure 8-2. Environmentally relevant committees exist in both houses of the U.S. Congress. At left are the committee names; at right are their areas of responsibility.

have roles to play in the passage and enforcement of environmental regulation. Their involvement is dictated by the very specific needs and conditions of a given area, as well as by the availability of local funding and proper authority. An argument can even be made that the effectiveness of certain environmental initiatives varies according to the type of initiative and the body implementing it (Weiss, p. 98).

Let us briefly examine the role of the U.S. Environmental Protection Agency, the one government entity that seems to have the greatest jurisdiction and most power in this arena.

The EPA: A Paper Tiger?

For all the grandeur of its name, the Environmental Protection Agency has but a limited reach in many instances. Part of the problem is geography.

According to Weiss, the EPA, like many other federal agencies, suffers from a lack of resources, personnel, and even expertise; as a result, it lacks the ability to administer environmental protection in every state (p. 102). Nor is there an EPA office in each of the 50 states. As of 1994, there were only 10 regional EPA offices, and some of those are located 100 or more miles away from civilization. Hence, because EPA officials are not physically close to, or familiar with, local environmental problems, they are not privy to the competing interests involved. Accountability does not fall to those satellite offices. Rather, it shoots straight to the top—to the EPA's Washington, D.C., headquarters.

Some states, even those whose environmental programs are derivatives of federal initiatives, may have no use for the EPA and decide instead to assume direct responsibility for the implementation of federally mandated environmental programs. State residents in these cases benefit from direct state delivery of services. There is, of course, a downside to shutting out the EPA, such as losing certain funds through nonparticipation in some federal programs—for example, superfund cleanups (Weiss, pp. 102–103). Also, Weiss writes, citizens may sue their home states for failing to adopt fundamental needs like groundwater protection.

Early on, the EPA had a large degree of discretion in implementing federal environmental protection programs. While this would seem to make the agency more powerful, in actuality that discretion opened the door to disagreement from affected bodies—who sometimes have gone on to file suit against the EPA. In some cases, Weiss adds, Congress has had to intervene when the EPA was hamstrung by these suits or its personnel simply could not decide how to respond to those challenging its interpretations.

One of the ways in which Congress has stepped in to save the day has been to write rules into environmental statutes. These rules made the statutes more specific. If the EPA failed to adopt or contest the additional provisions by a certain deadline, Congress's rules went into effect.

It could be argued that this kind of handholding by Congress may have the effect of weakening the EPA by taking away its autonomy with regard to enforcement and implementation. On the other hand, writing provisions that provide for state adoption of some programs—the Clean Air Act and the Safe Drinking Water Act, for example—is a way of recognizing that states are in a better position to adapt some laws to meet local needs. The flexibility of the EPA may be further curtailed by specific provisions in federal environmental laws.

And because, as we have indicated, state environmental programs are derived from federal provisions, states may also find themselves hampered as they decide how to adopt these federal programs (p. 104). Further complicating matters, writes Weiss, is that state programs "must meet specific national criteria before they can be authorized by the EPA" (p. 101).

Whether you believe that the EPA and the government as a whole should have wide environmental enforcement discretion depends upon your occupation, whether you are a business owner and the nature of your business, and where in the country you happen to live. The party in power may have something to do with the agency's effectiveness. Democrats, who generally favor more government control, are likely to earmark more resources for regulatory functions, while Republicans, who believe in less government interference, would be inclined to spend the country's money elsewhere.

The argument has been made that we have at times overreacted to environmental concerns.

But the EPA maintains a cautious interest in ISO 14000, declaring itself "officially neutral" on how it will treat companies that conform to the standards. The agency's chief interest is in compliance with laws, while the standard's focus is on complying with *management* dictates, leading an EPA official to call ISO 14000 insufficient as an enforcement tool. Another EPA official was quoted in a 1996 issue of *Environmental Science & Technology* as saying the agency was awaiting the results of a series of pilot projects that incorporate ISO 14000.

Mary McKiel of the EPA standards office said that while the agency may incorporate some ISO 14000 principles into its voluntary programs and criminal enforcement activities, it had no agencywide position as of October 1996. The EPA has more than 25 such voluntary programs, covering things such as toxic-emission reduction, reduction of methane emissions through manure management, and reducing greenhouse gas emissions.

"You can't say the agency has any official agency position," McKiel said.

Some of the EPA programs that may find a role for ISO 14000 are the Common Sense Initiative, the Environmental Leadership Program, the Consumer Labeling Initiative, and Project Excel, McKiel said.

"ISO 14001 is not a performance standard, and there are offices in the agency, or people in those offices, struggling with how in the world it can be useful if it doesn't have performance requirements," she added.

In any event, ISO 14001 is unlikely to replace the concept of external regulation, said an official at Arco Chemical.

"This doesn't replace regulation," the official said in *Environmental Science & Technology*. "What we're looking at is the method of achieving those standards."

Environmental Regulation and Economic Development—Mutually Exclusive?

Richard Stroup thinks that regulation and development compete. Stroup, the senior associate of the Political Research Center in Bozeman, Montana, has written that public pressure has often resulted in policies which are counterproductive to environmental policies, as well as health and safety policies.

In the 1990 book *Economy and the Environment: A Reconciliation*,* Stroup writes, "We are increasingly operating in a new regulatory climate" (p. 196) in which public and special-interest-group demand for a risk-free world has fed a bureaucracy that spends money to solve unproven problems and delays approval for potentially life-saving drugs because of even the most infinitesimal risk of cancer and other side effects. Stroup argues:

> In general, attempts to achieve "zero risk" in any single arena are dangerous to society because they stifle technical and economic progress. Most research and development efforts in biotechnology have shown no clear and present dangers and, indeed, offer the promise of more effective waste cleanups and less environmentally threatening means of controlling agricultural pests and weeds. But again, the atmosphere of fear and even of crisis has meant that research, testing, and use of these techniques has been hampered needlessly by a web of regulations created by the EPA. . . . The assumption that government controls aimed at reducing risk to zero are an effective way to make society safer is simply not true (pp. 197, 203).

He writes that the 1978 Love Canal incident has still not produced evidence of long-term health problems, and that a series of *governmental* errors led to chemical leakage at the waste site.

---

*Richard Stroup, in Walter E. Block, ed., Chemophobia and Activist Environmental Antidotes: Is the Cure More Deadly Than the Disease?, *Economy and the Environment: A Reconciliation*, Fraser Institute, Vancouver, British Columbia, Canada, 1990.

A few years later, in 1983, public outrage erupted over the case of Times Beach, Missouri, which was thought to have been contaminated with dioxin 10 years earlier. It is still unproven that anyone in the town was harmed by the poison, but the EPA spent $33 million to move residents. Despite then EPA administrator Anne Burford's personal appearance in Times Beach to announce the buyout, she was forced to resign shortly thereafter following allegations of "callous disregard" of the dangers of dioxin. Times Beach remains a ghost town despite the continued lack of evidence that there is any risk to humans (p. 194).

Overzealous government involvement in environmental issues, Stroup argues, has "catastrophic results" and poses serious dangers to our safety and material well-being (p. 192). In fact, the long-term risks of knee-jerk government action include stricter regulations, more taxing and spending, and large mandatory expenditures. A study conducted by the EPA itself concedes that its priorities are more likely the result of public opinion than of a genuine concern for public health. The corollary, of course, is that there is little incentive for environmental groups (which receive large sums from the government and panicked donors) to release information that might reduce public fear and uninformed outrage.

Stroup doesn't argue that those whose political and/or agency careers are tied up in chasing polluters are necessarily acting selfishly or in bad faith (p. 195). But their promulgation of public fear and outrage serves to expand their agency budgets and prolong their careers, making them one leg of what he calls the "Iron Triangle." Special environmental interests who use their lobbying power to help government agencies stay in business form the second leg, and politicians, who are often beholden to the special interests, are the third (p. 195).

The People: An Especially "Interested Party"

Public attention to environmental issues ebbs and flows as other interests and priorities come into focus in people's lives. Government, and candidates for government office, tend to respond accordingly in order to win and keep their jobs.

Let's take a moment to briefly examine just how strong these public sentiments are and what arouses them. This is important, as it reveals the intensity of environmental interest among those stakeholders whom you are seeking to satisfy through your EMS.

In "Environmental Policy in the 1990s,"* Christopher J. Bosso implies that vocal public concern over the environment tends to occur in two situations: when a particular occurrence puts the environment at the top of the political and media agenda, and when such concern does not run counter to personal priorities, such as maintaining employment (p. 32).

As an example of the first situation, the environment topped the news and the national agenda following such tragedies as the *Exxon Valdez* spill and the nuclear accident at Chernobyl, which is expected to have lasting effects. But these events occurred in the 1980s. In the 1990s, according to Bosso, came a "down period when the environment (took) a back seat to other, arguably more pressing, matters" (p. 32).

Bosso cites a January 1992 poll in which only 4 percent of respondents said that they wanted the presidential candidates to emphasize the environment in their campaigns (p. 32). This meant the environment was tied, in fifth place, with issues like drugs, the elderly, and abortion. Health care was first, at 19 percent. Four months later, only 11 percent of poll respondents listed the environment as the nation's most important issue, emphasizing the environment's fall from grace on the public's list of priorities (p. 32).

Then again, Bosso writes, environmental issues "never rank that highly" in polls that ask people to list the country's "most important issue." These polls usually cause respondents to tap into their own personal priorities, leaving the environment out in the cold unless, as we've mentioned, media attention to the environment and related issues is especially strong (p. 32).

"The polling data also suggest that citizens believe increasingly that a healthy environment is not necessarily antithetical to a strong economy. Only when environmental protection is linked to increased unemployment does its strong majority backing disappear," he writes, "and here the proportion willing to accept higher unemployment has grown considerably" (p. 32).

But the environment is no longer, at least in the 1990s, the darling of public opinion. There are limits, as Bosso acknowledges.

> Strong generalized support for environmental protection does not translate automatically into support for specific policies. It does not deter voters from rejecting ballot initiatives deemed overly ambitious or from siding with interests defending local industries or jobs. Environmental priorities do compete with other, often more salient

---

*Christopher J. Bosso, "Environmental Policy in the 1990s," *Congressional Quarterly*, 1994.

concerns during elections or policy debates. Even strong public concern does not translate into policy responses. It translates only into opportunities for leadership that may or may not be exploited (p. 32).

But politics isn't the only device for solving problems. What about well-known activist groups? Don't they present a united front in their single-minded passion for environmental protection?

Actually, these groups are no match for the development advocates, who have more clout and resources when it comes to lobbying Capitol Hill for their interests. Even if the activist group were single-minded (and as we shall see below, this isn't always the case), our fragmented political system diffuses power and makes it difficult for any one group or interest to have its own way (p. 34). When decisions are finally made, policymaking is such a filtered, tedious, and labored process that solutions to problems are rarely simple or balance all interests. For it to do so is virtually impossible. Like it or not, the dreaded "gridlock and drift" that President Bill Clinton excoriated in his 1992 inaugural address is a constant that can never be eradicated.

That makes two strikes against public environmental support: Not only are concerns over environmental protection tempered by personal economic worry, but the policymaking hierarchy is more hospitable to those with short-term, local interest and influence. These conditions muffled environmental voices in the 1990s, when such attitudes threatened the business and political status quo in a shaky economy.

Environmental Activism: Its Own Worst Enemy?

As with any social movement, the environmental groups started out as small entities that, with time, became well-entrenched, stable institutions (Bosso, p. 35). But what happens when several groups pushing the same cause must compete with one another for the two most valuable resources activist groups can have—publicity and money—to their own detriment?

By all accounts, public support for the environmental movement has skyrocketed since the 1970s, as evidenced by the severalfold increase in membership in specific groups such as the Sierra Club (one of the best known) and the National Audubon Society. Membership in the former increased four times from 1970 to 1990, from 150,000 to 600,000; the same pattern is visible with the latter (p. 36).

But as groups such as this grow into highly organized, structured, businesslike entities, with departments dedicated to different functions and even adjunct professional staff, the challenge for group leaders is "to manage ever-bigger and more expensive operations while holding onto the loyalty and commitment of a heterogeneous and often restive membership." These groups almost have to grow to survive, especially as the demands of being active players in the policy arena drive costs.

> But growth creates strains. Large memberships and big budgets combined often, if not always, generate pressures toward administration over advocacy, an overt focus on the budgetary effects of agendas and tactics, and a need to be seen as "respectable" so that potential contributors feel their money is spent wisely. Size also creates an emphasis on stricter and more centralized decisionmaking, often at the perceived expense of local chapters housing the group's most dedicated and least docile activists. Such friction in turn can generate damaging internal schisms, member defections, and the creation of splinter groups. This tug-of-war between national leaders and local activists, combined with the certain if muted willingness of other groups to recruit the disenchanted—only so many people are willing to pay membership dues—makes organizational maintenance an overwhelming burden (p. 37).

Bosso cites the Sierra Club as an example of such division (p. 37). Meanwhile, smaller groups without the desirable structural integrity of the larger ones—which lends visibility and credibility—must simply try to survive.

Sometimes the need for funding forces environmental activist groups to seek corporate support, a seeming compromise of their desired independence. Bosso writes that money continues to be an ongoing preoccupation of these groups, especially at times when recessions and national crises divert revenues to fighting for the cause of the moment.

For example, he writes, in 1990 these groups doubled their 1987 revenues, but shortly thereafter, because of the Persian Gulf war and for other reasons, the prominent National Wildlife Federation laid off staff, froze salaries, and left other jobs vacant because of sagging funding. The Sierra Club and Wilderness Society told similar tales (p. 38).

These financial crises sometimes have forced "wholesales" in recruiting tools such as organization magazines and other publications. Government and corporate donations are facilitated through "Earth Share," a conglomeration of the groups' fund-raising arms.

The still-unanswered question is, does such corporate funding buy influence? Bosso keenly observes that whether or not it does, what *is* true is that any threatened withdrawal of such support points up the danger in relying on corporate dollars (p. 38).

A good example is the National Audubon Society's public television series *World of Audubon*. The show was sponsored as far back as 1991 by General Electric, but faced the same dilemma as any television show that outrages a particular group upon whom sponsors rely. According to Bosso, GE (among others, namely Shroh's Brewery Co., Ford, and Citicorp) withdrew support from the telecast on the heels of threatened boycotts from an association that lobbies for cattlemen, loggers, and others who depend on the outdoors for a living. The association was highly critical of a *World of Audubon* series that disparaged government-funded cattle-grazing programs (p. 39). Whether this sort of withdrawal of support will be the trend remains to be seen.

As many outside observers would concede, however, when these groups work, they often work well. When it became widely publicized that a chemical growth regulator called Alar was used on apple crops, the Natural Resources Defense Council took its outrage directly to the people—over the heads of White House officials who were reportedly blocking action on the subject—by appearing on the CBS program *60 Minutes* to present their case against the apple growers. The chemical was subsequently withdrawn and its manufacture stopped. These results sent a message that was undeniably clear: Forget about "toothless" regulators and channels. Appeal directly to the public and let nature take its course (p. 40).

The other message was that environmental groups must risk their humble, spartan images in order to achieve their ends, even if their collective credibility then comes into question. Working *on* the government by directly influencing consumer choices became preferable to working *with* the government by lobbying special-interest-group-supported officials.

The latest brand of activist to emerge on the environmental scene is not necessarily associated with any group, large or small, well-known or obscure. We are talking about the New Environmentalists, which Bosso defines as "a grass-roots, community-based, democratic movement that differs radically from the conventional, mainstream American environmentalism, which always had a strong nondemocratic strain far too distant from real people and real problems."

One of the concerns voiced by such groups—and individuals—is the need for "environmental justice" that seeks redress of problems caused to specific regions or groups. One Midwestern university professor writes of the need for environmental justice in terms of removing incinerators and other environmental hazards from their common inner-city locations, where they are most likely to spawn disease and destitution among minorities and other underprivileged residents.

But he also calls for an overhaul in the "structural underpinnings" of the larger society, such as unemployment, that force certain segments of the population to become predictably trapped in surroundings where they will suffer from exposure to these and other societal ills—spousal and child abuse, alcoholism and drug dependency, suicide, despair, and overall destruction of their self-concepts.

Environmental activism, like other social movements, has seen victories and struggles. But activist groups clearly must rethink their organizational and administrative goals so as to avoid two things: losing ground to infighting and eventually losing sight of their original mission, and becoming overly dependent on corporate and government handouts.

For you as a manager, communicating with these stakeholders, whatever their internal tussles, and understanding their needs and wishes, is important.

9
Suppliers and Their Effects on ISO 14000

Maintaining Environmental Compliance within the Supplier Chain: A Matter of Common Sense

ISO 14001 registration tells the world that yours is a facility concerned about consistent and effective environmental management. But conducting business in this manner cannot occur in a vacuum, because you don't do business in a vacuum.

Depending upon the nature of your operation, you may have suppliers who, just like you, have some environmental impact. This is unlikely to be the case, of course, if you are a service organization or an information clearinghouse. But if you are a chemical manufacturer or oil company, you are part of a supplier chain, linked to individuals and companies whose operations could throw you out of ISO 14001 compliance if they don't *at the very least* adhere to some common sense environmental principles.

Your suppliers do not have to be ISO 14001–registered in order for you to give them your business. In fact, you have great latitude in deciding who is suitable and who is unsuitable, based on your facility's particular needs.

And there is no need to "convert the masses" so that they'll be willing to follow in your environmental footsteps.

"You can't force them short of transferring the contract," said consultant and trainer Ralph Grover. "You're the only person who decides where you'll spend your money. But you still have to buy that service or product."

Does this back you into a corner? No, says Grover. He said the effects of having no ISO 14001–registered suppliers in your chain "shouldn't be detrimental."

Ask yourself which of your suppliers have the highest environmental impact. Start your supplier-selection process by looking at them.

"You could be in an industry with virtually no environmental impact, such as a service company. Maybe you're buying paper and that's your principal product. If you're a chemical company and your suppliers supply you with raw chemicals, those chemicals are likely to have a significant impact," Grover said.

"You're going to want to know how those people are handling those chemicals—do they put them in a tanker car on a railroad that is not up to specifications? If you ship chemicals and you use a supplier who's a trucker, you want to make sure that trucker has the proper insurance, permits, training, safety procedures, the proper environmental management issues in place," he said.

This litany of questions is in sharp contrast to your business dealings with the "mom-and-pop" stationery store that buys from a wholesaler that buys from a manufacturer that buys from a forester.

"Worrying about the trees," Grover said, "is a lot longer a reach than worrying about chemicals. He's the guy next door."

He said that there is necessarily a certain amount of give-and-take and flexibility when choosing suppliers from among a pool where no one is ISO 14001–registered.

At the very least, he said, you should draw up a contract with a supplier that includes clauses requiring adherence to environmental regulations and other safety requirements.

"Any company that purchases products and services today should include, as a matter of good purchasing practice, reviews of regulatory compliance and safety," Grover added.

You also want to make sure that your suppliers are maintaining environmental permits and paying their taxes so that they're not at risk of being shut down by government regulatory agencies or the IRS.

Of course, there is always the possibility that you can convince a prospective supplier that it is in the supplier's best interest to become ISO 14001–registered, thereby creating a supplier chain. But you probably have other, more pressing day-to-day problems to consider. So it pays to compromise, Grover said.

"You can always get at some of these environmental management issues through purchasing agreements and bid selection," he said.

The Virtues of Minor Arm-Twisting

Another viewpoint on this issue suggests that minor arm-twisting can be effective.

"You as a customer decide what the requirements for your suppliers are," said Joseph Koretzky, environmental management systems facilitator for electronics manufacturer Sharp, recounting his own facility's screening process. The facility, in Washington state, manufactures LCD screens for laptop computers.

"What we did with ISO 9000 registration is, we said, there is a certain number of qualified suppliers. If suppliers were ISO 9000–registered, we accepted evidence that they were qualified. If not, we did a second-party audit in order to see if a supplier met our requirements.

"We'd see what management system was in place. If it was adequate, we would certify that supplier. I see the same methodology used for ISO 14000. In other words, if we feel it's important for our suppliers to have an environmental management plan, we're going to insist on it."

Should Koretzky come upon a suitable supplier who possesses all necessary qualifications except ISO 14000, he said, he will, according to a methodology developed by the facility, conduct a "pseudo-registration" of that supplier. This is followed by ongoing evaluation to make sure that the supplier is continuing to follow the precepts of good environmental management.

"This makes eminent sense," Koretzky said. "I wouldn't want some standards body saying, 'These are the standards your supplier must meet.' My priority is to define what the needs of my business are and to come up with a set of supplier requirements. It changes from company to company.

"We define a set of criteria," he added. "Suppliers must meet these to be approved suppliers. In addition, they must continue to perform at an acceptable level to be retained as an approved supplier. If they fall off the merry-go-round, we have steps they have to take to be recertified."

Down at the other end of the scale, Koretzky said, he has observed instances wherein ISO 14001–registration does not give a supplier an automatic advantage.

"I have found that some of our suppliers were ISO 14000–registered and their management system was lacking. We have basically not accepted their certificate as a prior judgment on whether they are certified suppliers.

"We have asked them to meet *our* management requirements," he said.

The amount of work you did to become ISO 14000–certified should not necessarily be expected from suppliers or prospective suppliers. Compromise should be the name of the game when rounding out your supplier chain.

10
Implementation: What It Takes

The Importance of Training

The watchword in ISO 14001 implementation is *planning*. An organization cannot simply start writing EMS documentation one day and decide that it has started to implement an EMS.

The good news is that companies are not up against any particular deadline regarding ISO 14001—unlike first-tier auto suppliers, who find themselves under increasing pressure to implement and register to QS-9000. So there is plenty of time to prepare by finding out what is involved.

Warren Bird, manager of the Chicago office of Arthur Anderson LLP, offered some preliminary investigative ideas in the September 1995 issue of *Chemical Engineering* (p. 96):

- Consult marketing staff members, who relate customer needs.
- Meet with operations personnel, particularly those who can estimate the costs and benefits associated with implementing ISO 14000.
- Involve ISO 9000 personnel (if applicable), whose program costs offer a good starting point for estimating the costs of complying with ISO 14000.
- Establish a primary objective calculation of the risks and rewards of moving forward. Include an analysis of a "do-nothing" scenario, since some of your firm's facilities may not need to seek ISO 14000 certification.

- Create a survey document to determine the number of facilities that could meet the standards today. It will improve cost-estimating accuracy.

Ask also how the standards can be applied within your company's existing culture.

Ask a member of your organization to perform research into ISO 14001 and the entire ISO 14000 series. This should be a person who is good at absorbing and digesting new information and presenting her or his findings objectively. The person should also be aware of the company's history of environmental compliance and the environmental backgrounds of its employees.

Send that employee to a two-day ISO 14000 overview course. The course will offer a detailed look at the core requirements of an ISO 14001 EMS, the process of implementation, and a discussion of registration requirements. Figure 10-1 gives an outline of the ISO 14000 overview offered by Perry Johnson, Inc., which includes a workbook entitled "ISO 14000 Overview: Preparation and Certification."

The company also offers an ISO 14000 Executive Overview booklet, free of charge.

Following the overview, ask your employee to present her or his findings to management. This presentation should answer the following questions:

- Is the company equipped to begin implementation at this time? Are necessary resources, such as time and knowledgeable employees, in place?
- What are the steps toward implementation?
- How long would it take to implement an EMS at a company this size?
- Is any special preparatory work involved? How soon can implementation start?
- How would the company benefit?
- What costs are involved?

Other general sources of information can supplement your employee's research. This book and others can be of great help. And calls to companies that have implemented an EMS and become registered to the standard (names of these companies are available from registrars) can be invaluable.

Begin your quest for ISO 14001 registration by selecting a group of individuals to spearhead the effort. How many people you choose and who they are will depend upon the size of your organization.

> **ISO 14000 Overview:
> Preparation & Certification**
> 2-Day Course
>
> I. **What is ISO 14000?**
>   A. Background of ISO 14000 environmental standards
>   B. Benefits of ISO 14000
> II. **Overview of Core Requirements**
>   A. Environmental Management Systems (EMS)
>   B. Environmental Labeling
>   C. Life-Cycle Assessment
>   D. Environmental Aspects in Product Standards
>   E. Environmental Auditing
>   F. Environmental Performance Evaluation
> III. **Implementing and Operating an Environmental Management System (EMS)**
>   A. Understanding The Requirements of ISO 14001
>   B. Assigning Roles and Responsibilities
>   C. Providing Training and Resources
>   D. Establishing an Environmental Policy
>   E. Setting Targets and Objectives
>   F. Developing Documentation
>   G. Establishing and Maintaining an EMS
> IV. **Monitoring the EMS through Audits**
>   A. ISO 14010, ISO 14011/1 and ISO 14012 Audit Standards
>   B. Roles of the Auditor and Lead Auditor
>   C. The Audit Process
>   D. The Audit Report
> V. **Meeting ISO 14001 Registration Requirements**

Figure 10-1. Perry Johnson, Inc., a multinational training and consulting firm, offers a two-day ISO 14000 overview.

Internal Auditor Training

Employees need to learn two things: how to implement an EMS and how to audit one. Not all employees, however, need to learn both. Long before you have completed implementation of an EMS, 1 out of every 20 employees in your organization should have taken an ISO 14001 internal auditor course and be prepared to conduct an audit.

An audit is an independent, planned, documented investigation of a quality management or environmental management system that is performed in order to determine how closely that system conforms to the standard that lays out the required elements of that system.

An internal or *first-party* EMS audit is an audit in which employees of an organization evaluate their own organization's EMS. Auditing yourself—taking a good, hard look at the results of your own implementation—is crucial, for it gives you the opportunity to correct any flaws in the system before undergoing a third-party registration audit by a registrar.

In April 1996, Perry Johnson, Inc.'s five-day Auditor of Environmental Management Systems Training earned the distinction of being the first such course to be accredited by the International Register of Certificated Auditors (IRCA) of Great Britain. Ralph Grover, one of the instructors, was an unfailing source of advice and counsel throughout the preparation of this book.

Figure 10-2 gives the outline of that indispensable course.

Employee-auditors, however, may not audit areas of a company in which they usually work. This rule ensures objectivity; an employee evaluating an area in which he or she works is likely to be biased, causing him or her to overlook nonconformities (deficiencies).

A *second-party* audit, which is not part of implementation or registration but should be defined nonetheless, is an audit in which an ISO 14001–registered firm (customer) evaluates a supplier in order to confirm that the supplier is also conforming to the standard (see Chap. 9). In cases where a customer makes registration a condition for doing business with its suppliers, these audits are a perfectly acceptable way to keep suppliers on their toes.

Second-party audits, which are not mandatory, may also occur after parties on both sides of the customer-supplier equation have completed the registration process.

Third-party audits are registration audits. They are always performed by bodies called registrars—bodies which have been accredited, or authorized, to grant or deny registration.

At least 5 percent of your workforce should be trained as internal auditors. A few of these individuals may also wish to take an ISO 9000 lead auditor course. The lead auditor coordinates the audit team and supervises the audit. Because ISO 9000 is the foundation of ISO 14000 and other standards, lead auditor training prepares a person to head up audits of other quality management systems within other organizations. A person must be certified as a lead auditor, however, in order to officially function in that capacity (see Chap. 2).

Implementation: What It Takes

ISO 14000 Environmental Auditor Training
5-Day Course

* This course is registered by the Governing Board of the International Register of Certificated Auditors (IRCA) and meets part of the training requirements for registration as an Environmental Auditor (EMS) under the IRCA Scheme (A10153).

I. **Making a "Down-to-Earth" Impact**
   A. Addressing Significant Environmental Impacts Through ISO 14000
   B. Benefits of Implementing an ISO 14001 Environmental Management System (EMS)

II. **Understanding ISO 14001**
   A. Overview of Core Requirements
      1. Environmental Management Systems (EMS)
      2. Environmental Labeling
      3. Life-Cycle Assessment
      4. Environmental Aspects in Product Standards
      5. Environmental Auditing
      6. Environmental Performance Evaluation
   B. Establishing Documentation
      1. EMS Manual
      2. Creating an Environmental Policy
      3. Creating Procedures

III. **Principles of Auditing**
   A. Internal Auditing Techniques
   B. Auditing Practices
   C. Roles and Responsibilities of Auditors and Lead Auditors
   D. Review of ISO 14010, ISO 14011/1 and ISO 14012 Environmental Auditing Guideline Documents.

IV. **Meeting Environmental Auditing Requirements**
   A. Understanding the ISO 14000 Audit Process
   B. Documentation Review
      1. Reviewing EMS Manuals and Related Documentation
   C. Preparing for an Audit
      1. Selecting and Preparing Your Auditors
      2. Creating Checklists/Working Documents
   D. Conducting Audits
   E. Corrective and Preventive Action
   F. Follow-Up

V. **ISO 14001 Registration Requirements**

Figure 10-2. This is the outline for the IRCA-accredited five-day ISO 14000 Environmental Auditor course given by Perry Johnson, Inc.

Educating some of your employees as auditors has another benefit: Once they have been trained as auditors, they will be equipped to look at your EMS with the objectivity of a registrar. They will understand how an audit takes place and be able to spot and facilitate the correction of any missing links in the system that would otherwise delay registration (also called certification).

Implementation Training

Employees should also take ISO 14001 implementation training. You will probably want to send more employees through this training than you will send through auditor training, because you will need fewer auditors than implementors. The task of implementing is also spread out over a longer period of time.

Implementation training should include to those members of your organization already responsible for environmental management, and a few individuals who perform environmental functions (changing filters, measuring and recording air emissions, etc.). Figure 10-3 explains why.

Another possibility, likely to be found in much larger corporations, is the recruiting of full-time in-house trainers who keep abreast of legislation and technical information in a particular area of concern and pass the information on to employees in specially scheduled sessions. They attend seminars, learn what they need to know, then return to perform the training.

At the outset of training, start with the obvious resources. Ask the employee who initially researched ISO 14001 for you to make a presentation to the company at large. If you are ISO 9000–registered, ask representatives of your ISO 9000 registrar whether they are equipped to answer your ISO 14000 questions. Also, look over trade publications for ISO 14000 information. Take the time to educate yourself.

One good way to decide who needs what training is to develop a training grid, such as that pictured in Fig. 10-4 which lists specific types of training and the number of employees who should receive it. This model is likely to be helpful in larger companies, where varied training needs are likely to exist for large numbers of employees (smaller companies may automatically choose to send all personnel through formal training).

Of course, budget restrictions may prevent you from fulfilling this training "wish list." But don't cut corners to the point of using training organizations or materials that don't seem entirely credible.

Implementation: What It Takes

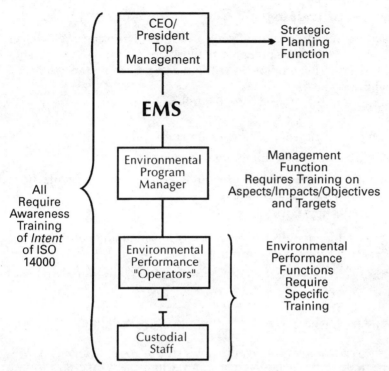

Figure 10-3. We suggest ISO 14000 training for those people who are responsible for environmental management in any given organization. This figure illustrates the division between various functions.

Sample Training Grid

TASK	# of Employees Who Need to Learn
Awareness	1100
Implementation	50
Internal Auditing	30
Documentation	20
Ongoing Technical (RCRA, HAZWOP)	20

Figure 10-4. You may choose to plot how many employees require training in any given area on a sample training grid such as this.

Approaches to Training

There is an infinite number of effective training approaches. Whichever you choose, plan well. Be prepared to work out cost and other logistics involved in gathering employees without compromising their full-time responsibilities.

Training should always address the following objectives:

- Instilling environmental awareness in employees
- Employing reputable training organizations
- Making the best use of time—find out whether training organizations offer convenient public seminars
- Expecting the best training
- Tying in training requirements with existing company policies for the sake of consistency and continuity
- Teaching employees how to respond to external environmental concerns

Screen any training company or individual using the following criteria:

Do their services meet your organization's specific needs? Can a firm provide training that accommodates your limited needs? If your training needs are broad, can you rely on this company to teach internal auditing, implementation, *and* lead auditing? Or will you have to jump between firms to get the necessary training, forcing your employees to endure a variety of teaching styles and philosophies?

Do the trainers know what they're talking about? If you have chosen to use a few especially astute employees to conduct in-house training, a major cost-cutting move, are these people really prepared to educate their coworkers? Developing a written test or putting these individuals through their paces with a mock training session may be a good way to find out.

Can training be customized? In the same way that television studios test their programs in front of small groups before exposing them to a national audience, you may want to "test market" a specific training program with a few employees, get their reaction to its effectiveness, then expose a revamped version to the whole of your organization.

An external training organization may be best equipped to determine how effective a particular training regimen was and make necessary adjustments.

What is the background of the trainer or training organization? Ask for references. Talk to the alumni of the organization's classes. What were the pass/fail ratios of these other students? Ask about the trainers' backgrounds and qualifications.

At the very least, trainers should be grounded in environmental technology and management systems. Regulatory and business knowledge, an awareness of overall environmental issues, and an understanding of ISO 14000's history and implementation issues all add to the value of a trainer.

Will the trainer supply advance outlines of their courses? This is a good way of determining whether a class will meet your specific needs. You may also want to ask whether the Environmental Auditors Registration Association (EARA) of the United Kingdom has accredited the course.

Is this a trainer with whom you can "grow"? The chances are that if you have had one satisfactory training experience with a company, you'll be satisfied following subsequent experiences. Hopefully, this company will be able to see your employees through ever-higher levels of training, progressing from overview courses through courses on consulting for specific implementation steps such as gap analyses.

You won't be training your employees to implement a system that tells the janitor what to do (no offense to janitors). Rather, you will be putting in place an EMS that sets forth the environmental responsibilities associated with that janitor's job, trains that janitor in those responsibilities, monitors performance, and documents the entire program.

Put another way, you're not concerned with changing the way Ken in waste disposal discards the plastic brackets that hold the picnic stemware together. Instead, you determine whether that activity can impact the environment (see below) and set about establishing procedures, to be carried out at the management level, that can alleviate those impacts. *You are not going to deal with traditional environmental compliance issues in ISO 14001.*

The training should give your employees a firm grasp of the steps to implementation. You may find it helpful to develop a Gantt chart as a way of plotting a timeline for the entire process. A Gantt chart is a bar graph whose x axis lists the necessary steps and whose y axis marks the amount of time each step should take.

Figure 10-5 shows a hypothetical Gantt chart, for illustration purposes only.

This training is expensive but worth it. Your employees will receive the most expert, up-to-date instruction available from people who are at

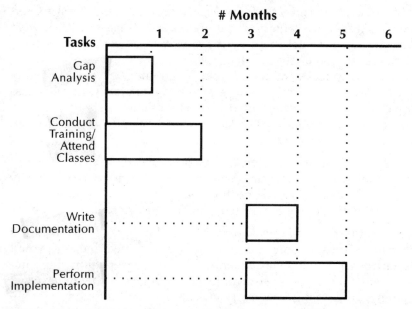

Figure 10-5. A hypothetical Gantt chart, which is a method of plotting how long various phases of implementation should take.

the top of this increasingly important field. For instance, Perry Johnson, Inc., recently presented on-site training to Ford Motor Company.

When the training is over, you can begin to conquer the EMS implementation process in a neat, orderly, step-by-step fashion. Here are those steps, in the form of an EMS implementation cycle (Fig. 10-6). How long each will take, and how difficult each will be, depends entirely upon the size and complexity of your particular organization.

Total Quality Environmental Management (TQEM): Benefits and Obstacles

Before we go further, let us examine another environmental management system methodology which subscribes to the doctrine of continual improvement: total quality environmental management (TQEM). Are ISO 14000 and TQEM compatible?

If you are focused on ISO 14001 and its very specific requirements and elements, it certainly may seem that you are doing all that you can reasonably be expected to do in order to achieve solid environmental man-

Implementation: What It Takes

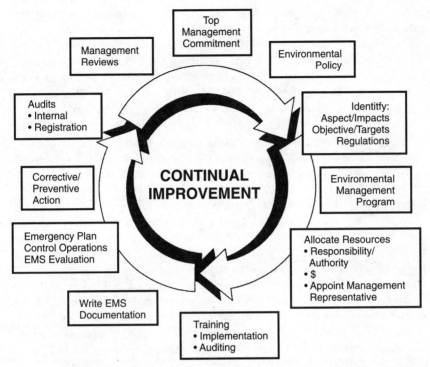

Figure 10-6. This cycle of continual improvement lists the major steps toward ISO 14001 implementation.

agement. But if you think of TQEM as a *philosophy* of continuous improvement that buttresses the practical commitment you've made via ISO 14000, then TQEM can work for you.

First, a quick review. TQEM, a specific derivative of TQM (total quality management), was the brainchild of the Global Environmental Management Initiative (GEMI) in its efforts to provide a methodical approach to the continuous improvement of environmental achievement. TQM is built upon four key elements, which coincide remarkably with some of the tenets of ISO 14000 (these elements also apply to TQEM): It is a management approach, it is long term, it involves everyone in the organization, and it is based on continuous improvement.

According to Steven J. Bennett, Richard Freierman, and Stephen George,* the tenets of TQM evolved just after World War II, when it was

---

*Steven J. Bennett, Richard Freierman, and Stephen George, *Corporate Realities and Environmental Truths*, Wiley, New York, 1993.

discovered that worrying about quality only at the end of the line resulted in declining quality.

The Japanese, say Bennett et al., suffered a similar fate during the war, when the country's focus was on producing the most powerful military machine possible. Because the civilian economy took a back seat to the war effort, quality on that front was "shoddy" (p. 33). After the war, recognizing that civilian goods were the backbone of the nation's survival, Japanese leaders sought advice from those people who had contributed to the quality of the products used by the U.S. military.

Hours of instruction by experts such as the notable W. Edwards Deming produced several strategies for revolutionizing quality. These strategies, focusing on management driving the process, continuous improvement, and everyone being involved, were the forerunners of what we know today as total quality management. Ironically, while it was the Japanese who sought American experts for advice on bolstering quality, American industry turned to Japan for techniques to improve the quality of its own manufacturing.

In applying this to environmental management, the third point, about involving everyone in the process, runs somewhat counter to ISO 14000 in that implementation is performed only by senior management and the results of that implementation are carried out by everyone whose job has an environmental component. Other than that, ISO 14000 and TQEM practically parallel each other.

Of course, no discussion of TQM is complete without a reference to the Malcolm Baldrige National Quality Improvement Act of 1987, created to award increased quality and productivity of U.S. companies. The criteria used by those who give this award have become a national model for the use of TQM in the United States.

It should be apparent that TQEM can be used as an ideological backdrop for the more structured procedures and requirements of ISO 14000. Like ISO 14000, TQEM is a systems approach which encourages environmental excellence through a proactive strategy. At the first-ever GEMI conference, chairman George Carpenter said that TQEM stressed the importance of pollution prevention and the improving of every aspect of environmental management.

"Total quality begins with accepting that we are never as good as we can be. Continuous improvement based on data and measurement is the fundamental bedrock of total quality," Carpenter said (p. 35).

Like ISO 14000, TQEM shifts the focus from compliance with regulatory and legislative requirements to customer, or stakeholder, satisfaction. Again, stakeholders are the surrounding community and world at large—in short, anyone affected by your operations. And these parties are increasingly aware of the fragile nature of the environment we all share and expect businesses to share that philosophy.

"In these times of growing environmental awareness, customers expect businesses to protect their health and preserve the environment," the authors write (p. 35).

Next, we'll explore how the pieces of TQEM fit together to form a well-oiled quality machine.

Management Must Take the Lead

The Baldrige Award criteria mandate that a company's leaders create a climate of motivation in which quality may be achieved. This is not a very involved process. It can mean as little as the circulation of a memo signed by top management which says something like: "As chairman and chief executive officer, I am dedicated to environmental excellence, and I plan to lend my support to these efforts. I expect all employees who wish success for this company to follow suit."

Conducting Environmental Training

Simply affirming a commitment to top environmental performance is not sufficient (p. 39) if it isn't accompanied by active participation in environmental management training. Bennett et al. write that management must take a hands-on approach by learning about applicable environmental issues and how these issues should be managed.

The management personnel who are best equipped to understand and implement these concepts are those who have statistical and problem-solving backgrounds.

Forming Environmental Steering Committees

This is the point at which management may begin to make employees aware of the company's environmental management goals and to impress upon those employees that they all equally responsible for reaching those goals.

The committee's responsibilities, generally speaking, are (p. 39)

- Providing direction for the company's environmental management process
- Establishing measures that the company can use to determine performance
- Monitoring progress in achieving environmental goals and adjusting the policy, goals, or course to keep the company on target

- Communicating goals as well as processes to achieve them and progress toward them to all employees
- Chairing cross-functional teams focused on companywide environmental issues

Corporate Realities and Environmental Truths says that the chairman and CEO of Eastman Kodak chairs that company's management committee on environmental responsibility. 3M went so far as to overhaul its entire corporate structure in order to facilitate the transfer of environmental decisions to the executive suite; environmental engineering is on the same level with other operating units.

TQEM Is a Long-Term Process

Because the goal, ideally, of TQEM is zero pollution, it is understandable that these are long-term processes. Pharmaceutical manufacturer Merck & Company developed a five-year plan that guides the entirety of that company's worldwide operations (p. 41).

Part of this strategy was to define the company's environmental policy and guarantee its acceptance and implementation. When that process ended in 1995, Merck drew up a list of lessons it had learned that could be applicable to other companies embarking on such an implementation (p. 42):

- Get attention at the top, since the process is led by management.
- Work overtime, but within a *set* and reasonable limit.
- Limit the number of planners. Less input means faster output.
- Promote communication among professionals in all areas—business, environmental, and health and safety. This type of exchange builds trust and understanding.
- Retain expert planners to facilitate the process, keep it moving, and keep it on track.

The Aim Is Customer/Stakeholder Satisfaction

As we have established, in environmental management terminology, customers are the surrounding community—everyone from your corporate and residential neighbors across the street to the world at large. Stakeholders are also local, state, and regulatory bodies—anyone who has a practical or legal interest in what you do and how you do it.

If conservationists are concerned about your use of plastic for detergent bottles, these concerns can be ameliorated by, as Downy did (p. 43), selling disposable refill containers so that customers can reuse their original plastic bottles.

Employee Involvement

No company can succeed at any venture as far-reaching as TQEM without bringing all employees into the fold. Otherwise, whatever management initiative is taken occurs in a vacuum.

Continuous Improvement

This is the final piece of the TQEM puzzle: extending any improvements that are made well into the future. We point to Procter & Gamble's approach to the continuous improvement cycle—plan, do, check, act—which it superimposes over its environmental systems and programs (p. 45):

> *Plan:* Procter & Gamble uses assessments, goal setting, and benchmarking to plan the elimination of pollution from its processes, products, and packaging.
>
> *Do:* Process control and pollution prevention are used to reduce environmental discharges.
>
> *Check:* Audits are used to analyze data and quantify results (as well as to understand both successes and failures).
>
> *Act:* Training is conducted in order to lay the groundwork for systems that sustain improvements and deploy site-appropriate systems.

As is the case with any management initiative, there is the danger of that initiative becoming the latest "flavor-of-the-month" program which is highly touted at first but eventually falls by the wayside or is overtaken by other priorities. Here are some of the possible obstacles to TQEM, followed by its by now obvious benefits (p. 46):

> *Lack of incentive to institute TQEM.* This is no small undertaking. If the incentive and motivation are missing, or there is no real corporate mindset aimed at TQEM implementation, it will fail.
>
> *Lack of management commitment.* If the CEO doesn't care, why should Joe Jones in shipping care either? Initiative of this type flows from the top down.

Short-term focus. This can't be reduced to a series of slogans and pep rallies. Start slow, with the focus on celebrating later.

Less-than-total employee involvement. Just as lack of management can influence the progress of TQEM, dissenting employees can have a similarly discouraging effect on the process.

Absence of reliable data and information. It helps if you can predict the outcome of the TQEM process—and if those predictions, based on scientific analysis and careful calculation, support the undertaking of TQEM in your organization.

A too-limiting definition of the word customers. Some management programs fail because their audience is too narrowly defined. Remember that the scope of your customer base in an environmental initiative is everybody—the government, neighbors, the world.

If these obstacles can be overcome, or avoided altogether, the return on your investment—in terms of improved customer/stakeholder satisfaction, organizational effectiveness, and better competitiveness—can be infinite. Planning is the key.

Back to our original question: Can ISO 14000 work in tandem with TQEM?

Clearly, the two concepts are highly compatible. The difference is that ISO 14000 is a specific system with requirements for audits, implementation of elements, and documentation. TQEM, on the other hand, is a philosophy that *may* be documented and methodical but needn't be—especially if you are being thorough about ISO 14000.

We suggest that in ISO 14000, you keep the focus where it belongs: top-down commitment, involvement of all employees (especially those whose functions are critical to your EMS), and improved and consistent environmental management. Improved performance, one of TQEM's main prongs, will be the byproduct.

Let's look next at how a company might undertake the implementation process.

Greener Grass Inc.: A Hypothetical Case in Point

This business and residential lawn maintenance company, headquartered in the 20,000-population Midwestern town of Sunnygrove, will undertake ISO 14001 registration. Company president Seymour Grubbs employs 27 drivers and 2 receptionists and maintains an office on-site.

Grubbs has decided that becoming registered will provide a competitive edge over other budding lawn maintenance companies who are moving in on his burgeoning customer base. He has no experience with the ISO standards, but he has read that implementing an ISO 14001 EMS is a widely accepted way to prove that one is using sound environmental management and is serious about environmental compliance issues.

Gap Analysis

Gap analysis is the process of deciding how wide a gulf exists between a company's current method of environmental management and that required by ISO 14001. Sometimes closing this gulf can require starting implementation from the beginning. In other cases, a facility may already be near the finish line.

At the lawn-care company, although the ISO 14001 standard has only six elements, Grubbs was overwhelmed. He was sure, after reading the standard, that he would have to close a lot of gaps during his implementation.

For example, he had only a skeleton of an environmental management system in place, a requirement of element 4.1. Because there was no environmental management system, there was also no environmental policy, required by element 4.2, to explain the management system or the company's commitment to continual improvement, prevention of pollution, or compliance with relevant environmental legislation.

No effort had been made in years to keep up with the changing regulations, a task that must be performed under element 4.3.2.

When information had to be communicated to employees, the most anyone had ever done was to pass around a memo, which usually ended up on the lunchroom floor or in the trash along with the fast-food wrappers. This is an absence of element 4.4.3.

Verbal means of communication had also been fruitless; drivers were too swamped to meet with Grubbs even for a few minutes to discuss new safety procedures, billing methods, etc. There was no way for Grubbs to figure out who knew what about updated company procedures.

Some elements of the standard—4.3.3, "Objectives and Targets"; 4.4.1, "Structure and Responsibility"; and 4.5.1, "Monitoring and Measurement" among them—require strict documentation, and Grubbs knew that he would have to start preparing this documentation and maintaining it in accordance with elements 4.4.5, "Document and Data Control," and 4.4.4, "Environmental Management System Documentation."

Grubbs was also worried about the lack of emergency preparedness and response, detailed in 4.4.7, especially considering the dangerous nature of the chemicals his drivers were working with, and the fact that they were stored at the company's warehouse adjacent to the offices.

Grubbs began filling these voids while he sent two drivers for ISO 14001 implementation training and two others for internal auditor training.

You needn't go this first crucial step alone. If you wish, you may hire a consultant to point the way. Some companies, Perry Johnson, Inc., among them, will even send a consultant to your site to facilitate a strategic planning session, or deliver an ISO 14000 overview (the firm also offers a free videotaped overview). If you choose to hire a consultant, however, you must do it in a manner that is prudent and judicious.

You must assume ownership of the entire registration process, all the way from this gap analysis through maintaining registration. Consultants may or may not be willing to do all the work, but this much is true: Whether or not you have learned anything about ISO 14001 or environmental management systems is of no consequence to them.

The entire scope of a consultant's concern should be solving critical technical issues and helping you get your implementation effort off the ground. The rest is up to you. If you choose to relinquish ownership of the process, even if you eventually attain registration, you will have done yourself a great disservice and you will be ill equipped to institute continual improvement or demonstrate a sincere commitment to sound environmental management.

Consultants must, above all else, be objective. They must not have a relationship with the registrar who will ultimately determine whether your EMS should be registered. If a company employs registrars and consultant/trainers, check out the structure of its finances. If the registrar end is run independently of the consulting end, you're OK even if the companies share a common owner.

Perry Johnson, Inc. and Perry Johnson Registrars are owned by one individual but have maintained this delicate balance for several years. You may find that philosophies on this subject vary from company to company.

If you are unconcerned about speed, take the time to send employees through the appropriate training.

If you decide to hire a consultant for your gap analysis, hire someone who has a grounding in environmental issue with strong management, particularly profit/loss, experience. An MBA holder is ideal. The idea is to avoid consultants whose forte is the nitty-gritty end of environmental performance.

Aside from their objectivity, consultants can be helpful for the following reasons:

A consultant will have specialized knowledge. Because consultants make their living specializing in a given area, they may bring a perspective and seasoning that helps you expedite your gap analysis or deal with another area you are having difficulty understanding.

A consultant will have seen many variations on the ISO 14001 theme. In other words, he or she has seen what works and what doesn't by having visited a number of companies like and unlike yours. So he or she may be able to help you cut a few corners and eliminate some obvious pitfalls unique to your operation.

A consultant will have credibility. A consultant has no reason to "make up things" about your organization. He or she is more believable in his or her assessments than someone who works there full time. This is a variation of the objectivity argument.

Consider these attributes of any consultant before bringing him or her into the fold:

- What are the person's credentials? Has the person evaluated a company similar to yours? Don't overlook lower-profile, solo-practice consultants because they don't carry the prestige of having worked for *Fortune* 500 companies.
- Who have been this person's previous clients? This is a question consultant candidates should be more than happy to answer.
- How knowledgeable is the person about ISO 14001?

If the candidate has worked with clients on this particular standard, all the better.

Better still is the consultant who participated on the United States Technical Advisory Group to Technical Committee (TC) 207, the committee that wrote the standard. Anyone who was in on this project is likely to have precisely the interpretation of the standard that was intended.

You should also consider personal chemistry:

- Does the candidate have a working knowledge of your field, or is he or she poorly prepared?
- Is the candidate a good listener, obtaining pertinent information before coming up with instant suggestions that are not applicable?

- Does the candidate do business cooperatively, allowing full participation, or does he or she simply pontificate and preempt discussion?
- Does the candidate understand your temporal and financial constraints?
- Is the candidate flexible, or are his or her suggestions of a one-size-fits-all nature?
- Is the candidate's knowledge of the standard thorough, or is it simply gleaned from newspaper stories?

However you choose to go about the gap analysis, do it with these questions in mind:

- Are there many gaps to fill? Will this be a major effort or a minor effort?
- How large are the gaps? Sometimes it's easier to develop an entirely new program from scratch than to make adjustments to an already-existing program.
- How will you fill these gaps? Have you decided what must be implemented?
- Are there any unanswered questions? How will you obtain essential information?
- Has management given the go-ahead to any and all required action? Will the process be slowed by having to get someone's approval every step of the way?
- Are some gaps more critical than others? Would it be easier to attack the small stuff first and then move on to the more complicated issues?
- Is there a deadline? How long, realistically, might this take?
- Who should be involved? What other resources are needed?
- How much will it cost? Is the money available? Has a budget been set?

Environmental Policy

Grubbs writes an environmental policy for his lawn-care business, perhaps one of the easier steps in implementation:

> Greener Grass, Inc., is committed to serving its customers' needs within a framework of environmental conscientiousness and attention to compliance. In transportation and application of lawn chemicals, personnel will handle the materials with care and respect. They

will stay abreast of the environmentally sound methods of storage and disposal of these substances, avoiding detrimental effects upon the community. Members of the public will have access to this policy.

Environmental Aspects and Impacts

Environmental *aspects* are those parts of an organization's operations which can cause a change in the environment and over which that organization has control. When these aspects actually *do* interact with the environment, especially in a negative way, they are called *significant impacts*.

The definition of "significant environmental impact" that most interests us is this one: a negative effect upon the environment that is severe enough to require legislative and regulatory action and/or may elicit angry public opinion is costly to the organization. This is what a company implementing an EMS wants to curtail.

In the case of Greener Grass, Inc., if a driver decides to dump excess lawn chemicals down a storm drain at the end of the day, rather than bring them back to the office for reuse, which adds an extra half-hour to his day, that is an *aspect* which causes a change in the quality of the environment.

Chemical contamination is a significant environmental *impact*. As part of his EMS implementation, Grubbs should seek to curtail the behavior of the rogue driver.

Legal and Regulatory Requirements

Staying on top of the changing face of environmental regulations is a key part of environmental compliance—and something Grubbs has pledged to do in his environmental policy.

Objectives and Targets

Objectives are broad goals, such as reducing the use of a particular chemical; the related target would be a 20 percent reduction by the end of the year.

Environmental Management Program (EMP)

The EMP is the device for achieving objectives and targets. Meeting each objective and target requires the allocation of resources, such as competent personnel and time, adequate to achieve the desired goal.

Let's stay with the chemical reduction example. If Grubbs chooses to use 20 percent less of chemical X, he must assign a person or persons to come up with a way to bring about this reduction within the desired time frame. Grubbs is essentially implementing a program within the overall EMS plan—a program that requires its own proper documentation and other elements.

Allocating Resources: Assigning Roles and Responsibility and Authority

Grubbs runs a relatively small company. The number of people he needs to include in implementation is small, because the nature, scale, and environmental impacts of the company's activities are quite limited.

He must still, however, as must you, appoint some kind of implementation team or steering committee. These may as well be the same few people he sent, during the off-season, through implementation and/or EMS auditor training (ideally he should take the training also).

When they return from training, they should be able to decide who will be responsible for what facets of implementation. Overseeing the entire effort should be someone appointed to the position of management representative.

Management Representative. This person should be selected from within the ranks of the company, but should not be the company president or other high executive. The reason for this is that management has responsibilities and obligations under the EMS which will eventually be audited by the third-party registrar. If Grubbs decides that he should be the management representative (MR), there will be no way to independently verify whether he has fulfilled his responsibilities under the standard.

Stanley Sprout, 33, is perhaps Grubbs's most trusted and loyal employee, the driver most likely to arrive early and leave late. He has often taken on additional responsibilities without being asked and has made it clear that he's ready for a new challenge.

Grubbs asks Sprout if he is prepared to take the position of MR and direct his coworkers—which is something he will have to face during implementation. Sprout will have no time to retain his driver duties during implementation. Being MR is a full-time job. Larger companies may have several M.R.'s.

In case there is resentment, the MR should feel no compulsion to justify his actions or decisions. He has the authority to act in the capacity of MR. Any disputes or questions should be directed to Grubbs.

Communicate Important EMS Information

Everyone in your organization should be made aware that changes are imminent, even if these changes don't radically affect every job in the organization. Avoid surprises.

By sending several employees for implementation training and internal auditor or lead auditor training, Grubbs has already let a large percentage of the employees in on his plans. They became instrumental in putting these plans into effect early in the process. If you haven't already let other employees know that their cooperation will be invaluable, now is the time to do so.

Prepare Documentation

It may be easier to write your EMS documentation if you think of the project as a pyramid, with each type of documentation flowing from the one before it. Perry Johnson, Inc. uses the configuration illustrated in Fig. 10-7.

Some key documents include the following:

Environmental policy. As we've said, this policy, which is mandatory, describes the organization's commitment to sound environmental management, its commitment to continual improvement, and a pledge to observe environmental regulations.

EMS manual. We frankly don't understand why anyone would go to the trouble of implementing an EMS without some master reference document that explains the system and makes reference to other pertinent documents.

The manual *is* a requirement of management quality systems, such as ISO 9000 and QS-9000, and may become a requirement of ISO 14001 in the future. At this writing (December 1996), it isn't.

Write one anyway. Use it to:

- Give background on the organization, such as a company profile.
- Reiterate your environmental policy and your commitment to continual improvement.
- Delineate responsibility and authority for the various components of your EMS.
- Describe the structure of your EMS.
- Address the organization's history of environmental compliance and reputation.
- Demonstrate management's commitment to the system.

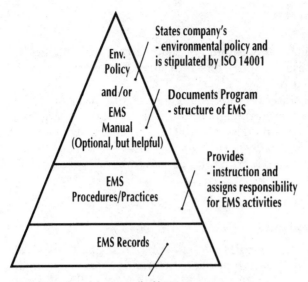

Figure 10-7. The ISO 14001 documentation pyramid. Environmental management system (EMS) documentation begins with the environmental policy, which is reiterated in the EMS manual, and ends with EMS records, which prove the existence of your system.

- Direct the reader to other documents.

Include a distribution list. That is, list all personnel who should receive and maintain a current copy of the manual. Not all employees need to hold it—only those who are instrumental in running the EMS. A sample EMS manual is included in Chap. 12. If you take a quick look at it now, you will see that it needn't be a literary gem. A concise 10 to 30 pages, depending upon the size of the facility, are sufficient. It is, after all, merely a reference tool. A great deal of the manual should direct the reader to already-existing policies, organizational charts, etc. The manual provides references to the subordinate documents in the pyramid.

EMS procedures/practices. This portion of the pyramid should provide instruction and assign responsibility for EMS activities. It should state the functions within the organization that make up the EMS and who is responsible for them.

For example:

- Monitoring effectiveness of EMS: Jerry Smith
- Updating company on environmental regulations: Mary Jones
- Controlling documentation: Billy Kart

EMS Records. These are the documents that support the EMS and prove its existence. Many different documents fall under the category of "Records," and if you expand your thinking, you'll see just how wide that range is:

- Descriptions of the company's environmental aspects and impacts
- Plans for corrective and preventive action
- Plans for continual improvement
- Documents supporting the environmental management program
- Forms, files, reports
- Drawings and specifications
- Regulations
- Expense reports
- Test information
- Material review results
- Documentation of complaints and remedial actions taken
- Results of internal audits

You need not perform an internal audit before you generate EMS records. These records can even be whatever notes you took in the planning stages of your EMS, because those notes might be handy when you're trying to think back to something you decided early in this process.

Any and all documents you generate during the creation, implementation, and evaluation of your EMS should be retained and organized as EMS records. They should be clean, dated, easily comprehended, and readily identified. Keep them out of harm's way.

Grubbs put one of his secretaries in charge of maintaining and organizing the records. She must verify that all the above conditions are met before a record is declared valid and filed.

An important distinction should be made between *records* and *documents*. Because a document is any piece of paper or computer file used in the preparation, explanation, or verification of something, records can also be called documents.

Documents, however, serve a far wider range of purposes than verifying things, and so not all documents are records. A document that lays out training plans is not a record. Nor is a communication procedure.

Records always = documents

Documents don't always = records

Records are the documents which though not currently in use support your EMS. Documents which are required by elements of the standard are referenced in the EMS manual.

If you decide to hire an outside consultant or firm to write your EMS documentation, you will be charged according to how many days it takes to complete the job. Consultants are required by convention to use the number of employees in your company as a guide in determining how many days are appropriate. How much they charge is up to them; how this figure is computed is not.

The international standards community, at this writing, had not developed a timetable for the writing of ISO 14001 documentation.

Develop Control Operations

This step refers to development of plans on how to respond to parts of the organization's operation that may cause unforeseen significant environmental impacts. This step is better explained by combining it with the next one.

Establish Emergency Response Plans

You must decide how you will respond if an accident or some other emergency situation causes an environmental impact.

Grubbs might decide that upgrading the chemical tanks on his truck fleet is a good way to avoid spills which might end up dripping down nearby storm drains and contaminating drinking water. This is a preemptive control operation.

His emergency response plan is to equip each driver with the telephone number of Sunnygrove's Department of Public Works so that that office can be called and have the opportunity to respond quickly in the event of a spill.

Procedures should be in place that facilitate the *periodic* monitoring of operations that may cause significant environmental impacts, as well as correction of nonconformities.

Management Review

In addition to audits, periodic management-level reviews of the EMS should be carried out in order to determine whether the system is adequate, operating as planned, and to ensure its continued suitability and effectiveness in light of changing conditions and requirements.

Grubbs asks Sprout to conduct an independent review of the EMS annually and report his findings during the management review. If no problems are discovered, Sprout may simply tell Grubbs that all is well; he needn't burden Grubbs with excessive details unless Grubbs wishes active involvement.

How long all this takes depends largely on your organization's size and complexity, as well as the diligence of those involved in the implementation. It also depends upon how much of an EMS you already have in place and how many of the elements of ISO 14001 you saw fit to adopt. It's likely that you just happened to be conforming to some of these elements all along.

You know you have an EMS in place when all documentation is in place and the requirements of the standard are being followed. You may have to take it on faith that you've completed all the steps to implementation. No lights will flash and no bells will sound to tell you that the job is done.

Grubbs took six months of intensive work in order to achieve ISO 14001 implementation, not counting the time he and his employees spent in training. But because an organization is constantly updating and improving, you're never really done with an implementation.

What Are Your Environmental Costs?

We have focused throughout this book on the ISO 14000 series as it applies to manufacturing companies. This may seem narrow, because any company of any size, function, or level of complexity may implement and register to the standard. Even a small service company with no design, manufacturing, or distribution functions may register.

For the sake of discussion, we will continue to consider ISO 14000 with an emphasis on production as we look at the concept of costing—determining the costs and benefits of corporate activities related to the environment.

The specific costs will depend upon your business and your accounting system, but one thing is universal, and it's not addressed in the standard: An adequate system of identifying and measuring environmental costs is essential when designing and implementing any environmental management strategy, according to the book *Measuring Corporate Environmental Performance.** The book says that without such a system, effective environmental management is "impossible" (p. 145).

And most companies lack such a methodology (p. 145). They have their work cut out for them, because this process can be complicated by the fact that future costs stemming from present-day products and processes are often not factored into the costs of those present-day processes and activities. Put another way, they're not preparing to pay tomorrow for what might happen today. Further, environmental laws function in such a way that future regulatory costs may be applied retroactively—a double hit.

Today's here-and-now costs may be difficult enough to understand and account for, without the added burden of having to shift your financial focus between the present and the future (pp. 145–146).

Accounting systems are beginning to change (p. 146) so that companies are better equipped to deal with this dilemma. The cost in blood, sweat, and tears to find an appropriate system may be considerable, but many companies are beginning to utilize systems that accurately measure past, present, and future production-related environmental costs.

ISO 14001 may be quite specific about document control, operational control, communication requirements, and the like, but its evaluation standards for products—ISO 14020, 14040, etc.—don't tell you how to determine what you'll pay to abide by its labeling, life-cycle assessment, and other requirements. The prescribed changes in product labeling, verifying environmental claims, and life-cycle assessment don't come about overnight. How you go about implementing the standards is up to you. You will have to work very hard to learn how to factor the cost of it all into your accounting.

According to *Measuring Corporate Environmental Performance,* you will deal with three general categories of costs (p. 146): current costs for past sins, current costs for current sins, and future costs for current sins.

As to the first, some companies already bear substantial cleanup costs for pollution that occurred several years ago. Some of these costs are related to the EPA's Superfund projects, the cleanup of which is now governed by strict and often retroactive legislation. When these substantial liabilities are added to the costs of manufacturing, the profitability of a company's operations (products, facilities, divisions, etc.) is affected.

Some companies, the book says, are aware of this and are dealing with its practical consequences. Naturally, controversy arises when profitabil-

---

*Marc J. Epstein, *Measuring Corporate Environmental Performance: Best Practices for Costing and Managing an Effective Environmental Strategy,* Irwin Professional Publishing, Chicago, 1996.

ity is reduced by these costs. In some cases management suffers, because, when the bottom line is depressed—even artificially—their performance evaluations and resulting compensation may suffer (p. 147).

There are, of course, other ways to distribute these costs. They can be included in company overhead, general administrative costs, shareholders' equity, or an account that is unrelated to production. This would result in overstated prior earnings, but current income would be left undistorted. This way, current costs for past sins needn't be identified as product costs or, for that matter, costs related to the current or past performance of any specific department. Employee reviews would be unaffected.

Whatever the decided-upon methodology, the controversy points up the lack of planning for future environmental impacts and failure to implement life-cycle assessment (p. 147).

On the other hand, not only is factoring present-day environmental costs into present-day activities uncontroversial, it's just good business sense. Even so, not all companies engage in this practice by separating environmental costs from other costs. Hence, exact product costs are determined inaccurately—leading to surprises later. And when costs are assigned and determined arbitrarily, undercosting and overcosting may occur. Later analysis is difficult because exactly which products caused the company to incur which environmental costs remains unknown.

It is generally agreed that disregarding possible future costs for current sins (the third category) is what leads to the first category, how to pay today for what happened in the past. Some argue that future costs are difficult to determine and should be ignored, but cost estimation models are available today that allow a company to better prepare for the future. One cost of ignoring consequences that we have already alluded to is that involving past Superfund cleanup of projects, which is still being paid for. Current costs must include an estimate of *total* product costs in all categories in order to relieve future managers and executives of sanctions over costs you have caused the firm to incur.

When truly pressed, companies which have previously described the estimation of future costs as impossible should be able to come up with some estimate (p. 148).

Ignoring the issue of costing is clearly improper (p. 149) and just plain harmful. The many companies that have yet to commence the practice of costing are likely to be in for some unpleasant surprises—and, aside from company profits being eroded, unsuspecting managers will be punished.

Epstein admonishes: "Improved costing systems must be installed in organizations to properly identify, track, measure and manage environmental costs."

11
ISO 9000: The Forerunner of ISO 14000

History of ISO 9000

ISO 14000 and other international management system standards owe their existence to ISO 9000, the first such series of standards developed by the International Organization for Standardization (ISO) in Geneva, Switzerland.

Let us look briefly at ISO 9000 and how it came to serve as the forerunner for ISO 14000.

ISO 9000, a quality management system standard, has nothing to do with the environment, the automotive industry, or any other specific type of commerce. It is a generic, highly adaptable set of 20 steps, or elements, which can make a company more efficient and quality-oriented.

The ISO 9000 series was born of a desire to break down barriers to international commercial traffic by eliminating quality requirements unique to various countries. Similarly, ISO 14000 was intended to eliminate conflicting environmental requirements as a trade barrier.

The idea behind ISO 9000 was that if all countries can agree on a universal way to test products, monitor the quality of customer-supplied products, and control the design process, the commercial output of one nation is likely to satisfy customers in any other nation.

Other business practices addressed by ISO 9000 include management responsibility, quality planning, contract review, organizational and technical interfaces, design input and output, document and data control, purchasing, and inspection and testing. Guidelines are also in place for auditing a management system to determine how closely it conforms to these requirements.

Figure 11-1 lists the 20 elements.

At this writing, adoption of neither ISO 9001 nor ISO 14001 was required by any legal or business mandate in the United States, although both have become a way of corporate life overseas. The European Union (EU) has been quick to adopt ISO standards out of a desire to unify its 15 member nations into a single marketplace in which goods, services, and capital can move freely from one country to another.

The ISO 9001 Standard

4.1	Management Responsibility
4.2	Quality System
4.3	Contract Review
4.4	Design Control
4.5	Document and Data Control
4.6	Purchasing
4.7	Control of Customer-Supplied Product
4.8	Product Identification and Traceability
4.9	Process Control
4.10	Inspection and Testing
4.11	Control of Inspection, Measuring and Test Equipment
4.12	Inspection and Test Status
4.13	Control of Nonconforming Product
4.14	Corrective and Preventive Action
4.15	Handling, Storage, Packaging, Preservation and Delivery
4.16	Control of Quality Records
4.17	Internal Quality Audits
4.18	Training
4.19	Servicing
4.20	Statistical Techniques

Figure 11-1. The 20 elements of ISO 9001.

Relationship between ISO 14000 and ISO 9000

Because of its almost universal applicability and flexibility, ISO 9000 has served as the basis for the equally general ISO 14000, which has emerged on the international scene and garnered wide acceptance.

The subject of our discussion in this book, the ISO 14000 series, contains the standard for the formation of an environmental management system (EMS). ISO 14001 is a management standard applied to a company's environmental aspects, as opposed to a management standard applied to quality in manufacturing.

ISO 14001 neither recommends nor mandates limits on air emissions, discharges into water, etc. You decide which of your company's environmental effects need to be addressed; the standard tells you how to implement a system which addresses them.

(ISO 14000 is the series of environmental management documents and ISO 14001 is the actual standard, as is the case with ISO 9000 and 9001.)

ISO 14000 is most closely allied with ISO 9000 in that both series are

- Voluntary
- Management-oriented and do not establish required performance levels
- Designed for internal use, although registration is by third-party assessment
- Generic and applicable to virtually any business of any size

  They are different, however, in that

- ISO 9000 is a quality management series, whereas ISO 14000 deals with environmental management.
- ISO 9001 requires a quality manual; ISO 14001 does not.
- ISO 9000 satisfies the needs and expectations of commercial customers; ISO 14000 is aimed at a much larger audience.

As ISO 14001 becomes a recognized, accepted standard, a properly implemented environmental management system may satisfy regulatory agencies, the public, and your customers that you are paying attention to the environment and the effects you have on it.

Reaping the benefits of a standard such as ISO 14001 is accomplished by "registering" to it—proving to a third-party auditor called a registrar that you have established, implemented, and maintained an environmental management system in conformance with the applicable elements of the standard. Registration may be driven by several factors, such as the need to compete with other firms which have become registered and a desire for access to overseas markets where registration is a prerequisite to doing business.

Many companies are likely to have become registered to ISO 9000 by now. If you are one of them, and you wish ISO 14000 registration, have you put yourself at an advantage? Or, do you have another huge amount of work ahead of you?

"I don't like the word *huge*," said EMS consultant/trainer Ralph Grover, who offered a slightly more detailed analysis of the differences and similarities between the two standards.

Grover said that the transition between ISO 9000 and ISO 14000 needn't be difficult, depending upon the company's size, function, product manufactured, complexity, number of facilities, and, of course, environmental impact(s). [ISO 14000 registration is by *facility*, not by company, because the environmental impacts of each of a company's facilities are different (e.g., headquarters vs. research and development lab)].

In fact, simultaneous audits for the two standards are possible, even though the two standards are not tightly harmonized. Delegates from the committees that composed the two standards were working in 1996 on harmonizing them.*

"The best guess," comments Grover, "if you're ISO 9000-registered, assuming a reasonable-sized company and a manufacturing process where you have a 'normal' amount of permits and regulations, is six months to a year to get registered to ISO 14000. If you're a small service organization and you're 9000-registered and want ISO 14000, it's not going to take you that long at all."

There are some fundamental similarities as well as some striking differences which must be taken into account.

"The primary common points, not exact," he said, "are the audit functions. That's where they are most similar.

"Otherwise," Grover said, "they're written for two different purposes. One is written to a quality system and the other is to a management system. They're both management tools; one is for quality and one is for the environment.

"The difference between ISO 9000 and ISO 14000 is fundamental, and, because of this fundamental difference, in that they each describe a dif-

---

International Environmental Systems Update, May 1996, pp. 8–9.

ferent type of activity within the company, the systems that are going to govern them are going to be different," he said. "The records and documentation are going to be different, in the sense that you have within ISO 9000 and ISO 14000 a document system and a records system that describe different aspects of the facility's business. You could keep ISO 9000 and ISO 14000 documentation together, but from a practical standpoint—administration and auditing—it makes it simpler if they are segregated."

Admittedly, ISO 9000 registration can be a great framework for ISO 14000 implementation and eventual registration. If you have been through one implementation and subsequent audit, you are already familiar with what will happen under ISO 14000. But you must focus your thinking on another "layer" of the facility.

"You may be involving personnel who have dealt with whatever management system ISO 9000 describes. Here, you will be doing similar activities, but you will involve different subjects and personnel. If you involve the *same* personnel, you're going to have to make sure they understand the difference between what their job is under ISO 9000 and what it is under ISO 14000," Grover said.

Whether you employ a consultant depends upon whether you think you can find a dedicated insider to do the job for you.

Joseph Koretsky, EMS facilitator for Sharp, the electronics manufacturer, says that a facility *without* ISO 9000 in place will have about "three times more work" to do, since ISO 9000 lays a basic quality infrastructure which you will be lacking.

Koretsky also said that the activities—document control, internal audits, management review, corrective action, and calibration—are the same for the two standards, but they must be performed differently.

"The amount of work we had to do was rather small," Koretsky said of his facility, in Washington state, which manufactures LCD screens for the monitors of laptop computers. He said the probability of any company going for ISO 14000 without first implementing an ISO 9000 quality system is slight because going for ISO 9000 first is "the right thing to do."

"If you're not ISO 9000 but want to go to ISO 14000, that's OK, but you've got to understand that you're going to make a huge investment."

On the other hand, he said, the incremental expenses involved in making the 9000 to 14000 leap are "not that great." So consider getting ISO 9000 out of the way first.

"Quality management is an integral part of business now; environmental management is not. But if I had to look at the two, I believe the potential benefit from implementation of ISO 14000 is greater than from ISO 9000," Koretsky said. "The ramifications in terms of the effect on the world are greater with ISO 14000 than with ISO 9000 because you're not affecting only quality; you're affecting the whole world—your neighbors, cities, and municipalities in general are being affected by what you do."

12
A Sample EMS Manual

Why an EMS Manual Is Necessary

ISO 14000 does not require that you write a management system manual. But we believe that, as is the case with other prominent standards, a manual is desirable for optimum system performance. It mystifies us why the creators of ISO 14000 chose to leave this requirement out of the specification standard, but since you are certainly free to write one anyway, we strongly recommend that you do so.

First, get rid of any mental images of overweight, overwritten, and verbose quality and policy manuals that no one reads or would want to read. We are talking here about a short, capsulized document that *refers* the reader to all relevant information rather than supplying it. We are talking about a summary of your ISO 14001 EMS and simple statements that affirm the existence of all its core documents.

Your EMS manual can be as short as 15 to 20 pages. In fact, the documentation pyramid that we recommend is based on keeping the EMS description short and simple and leaving other types of information, such as procedures, to another tier.

To simplify even further, all you are doing in an EMS manual is saying that all elements are in place.

The ISO 9000 Quality Manual as an Example

Before we show you how to write an EMS manual and offer a sample, let's learn more about ISO quality manuals using the other major ISO standard series that exist—ISO 9000, which does require a manual. The following is taken from *How to Write Your ISO 9000 Quality Manual,*© Perry Johnson, Inc.

> In ISO 9000 terms, the quality manual's chief aim is to prove that the facility's quality system meets the standard. While on-site assessments are part of the ISO 9000 registration process, the quality manual is the initial and, arguably, the most critical element of the process. It is not enough to have a quality system which conforms to the standard; the facility must also have a quality manual which satisfies the ISO 9001 standard's requirements.
>
> The characteristics of the ISO 9000 quality manual will surprise many who are acquainted with "traditional" U.S. quality manuals. The chief difference is length. ISO 9000 quality manuals usually run no more than 50 pages, even for the largest and most involved facilities. This is because the ISO 9000 quality manual avoids detail. Rather, it serves as a general guide to the facility's quality system. It covers major issues, addresses all components of the related ISO 9000 quality system standard, and provides cross-references to lower levels of documentation, which may include operating procedures and work instructions.
>
> Another big difference between ISO 9000 quality manuals and others is their relationship with reality. Unfortunately, many corporate quality manuals are management "wish lists." Too often, such manuals expound theory but signify nothing about the firm's actual quality system or lack of one. An acceptable ISO 9000 quality manual cannot be merely an expression of what management wants the quality system to be. It must detail precisely what the quality system is in practice. Since any variance between the quality manual and actual practice will be detected during initial or continuing on-site surveillance assessments, the manual must be completely factual.

Steps in Writing an EMS Manual

Let's examine the basics of how to write an effective ISO 14000 EMS manual. There are four steps to this process:

1. Learning what the standard requires

2. Determining the degree to which your facility's environmental management system conforms to the ISO 14001 standard
3. Identifying the specifics of that conformance
4. Expressing the specifics of that conformance in text

Elsewhere in this book, we discuss the standard's requirements, determining your current level of conformance (gap analysis), and identifying the specifics of that conformance. In this chapter, we will demonstrate how to fulfill the last step.

As we have pointed out, the standard has six elements, a few of which have subelements that contain more requirements. A quality manual, and hence an EMS manual, accounts for how your facility fulfills requirements. The best way to write a manual is to devote one page to each element (or subelement) that applies. Remember, you may already be in conformance with some elements.

First, decide who will be responsible for composing the manual. It is important that one person or team of persons take on and complete the whole project, because that ensures consistency and accountability. The writing team, which must be chosen by management, should have a good working knowledge of the part of the process they'll be documenting, understand the standard, and be genuinely motivated by a desire for sound environmental management.

Your manual will be easier to write if you have a smaller facility. Smaller facilities have fewer people and fewer processes. Either way, small or large, your facility will be mirrored in your EMS manual.

As you begin to write, remember that the EMS manual is the most general of all documents in the pyramid. To illustrate, we offer this side-by-side comparison of the EMS manual and all subordinate documentation.

We will say it once more. The EMS manual merely states that you are in conformance with all applicable elements of the standard.

How to Start

Plan to include the following elements:

Title, scope, area of application

Table of contents

Comparison of the EMS Manual and Subordinate Documentation

The EMS manual	Subordinate documents (procedures, records, etc.)
90 percent general, 10 percent specific	90 percent specific, 10 percent general
Brief and to the point	As detailed as necessary
90 percent principles and objectives, 10 percent "how-to"	90 percent "how-to," 10 percent principles and objectives
Addresses standard's elements point by point	Amplifies general statements according to procedural and other requirements
Changes less frequently	Changes more frequently, as technologies and processes used change
Refers to subordinate documents	Refers to the environmental management system manual

Introduction, containing information about the organization and the manual itself

EMS policy

Brief description of the organization's functions, defining responsibility and authority (as appropriate)

Section of definitions (if necessary)

Description of elements of the EMS and *references* to procedures

Usage guide/circulation list

You might include a "header" and "footer" on each page with some boilerplate title information, the subject of the section, and page numbers (1 of 1, 1 of 2, etc.). Here's what the typical header would look like (you are free to decide on your own format, but there's no need to get fancy; the EMS manual is by definition a spartan document):

EMS Manual	Kleen-Air Inc.	**Element 4.2** Environmental Policy

All necessary introductory information is here, and it is handy for anyone flipping through the manual in search of a specific section. The reader merely needs to look at the header for quick identification of the material covered on a given page.

The footer portion of the page contains an approval signature, revision numbers, and dates of approval. Sometimes more than one person will need to approve a section. Because the manual is not likely to change frequently, you will have few revisions.

Written by:	Name	Revision Number:
Approved by:	Signature:	Date:

Between these two standard pieces, you will sandwich all the information we have listed above. You may choose from a variety of ways of expressing the information, especially your conformance with the elements of the standard. Some people like to use a numbering system like this one:

3.1 Purpose (listing what the section covers)

3.2 Scope (listing what part of the facility is affected)

3.3 Responsibility (listing who, by title or department, carries out the function)

3.4 References (to supporting documentation and the applicable element of the standard)

3.5 Definitions (of terms unique to the manual and company)

A Sample EMS Manual

With that, let's look at the EMS manual for the fictitious company Kleen-Air Inc., a waste collection and disposal company.

| EMS Manual | Kleen-Air Inc. | **Title, Scope, Application** |

This is the Environmental Management System (EMS) Manual for Kleen Air Inc., an incineration facility in Anytown, USA. The manual applies to all aspects of the company's operation, from separating waste to minimizing the surrounding community's exposure to hazardous incineration byproducts. We also prepare waste that is unsuitable for incineration for transport to landfills.

Written by:	Name	Revision Number:
Approved by:	Signature:	Date:

EMS Manual	Kleen-Air	Table of Contents

Topic	Page Number
Title, scope, and application	1
Table of contents	2
Introduction	3
EMS policy	4
Description of function, responsibilities	5
Definitions	6
Description of conformance to ISO 14001	7
Usage guide/circulation list	15

Written by:	Name	Revision Number:
Approved by:	Signature:	Date:

EMS Manual	Kleen-Air Inc.	Introduction to organization and manual

A Sample EMS Manual

Kleen-Air Inc. was founded in 1967 as a subsidiary of Waste-Ed, a trash collection and disposal firm that contracted with school districts and municipalities and that still exists. Kleen-Air operates an incinerator at 139 W. Howe, Anytown, that operates 24 hours a day and burns trash from the above-mentioned school districts and cities (350 to be exact). Recyclable materials are not incinerated.

Kleen-Air employs 250 people and operates at a 25,000-square-foot facility bordered by the cities of Brown and Niles. The trash we burn is collected within a 500-square-mile radius.

This EMS manual describes the environmental management system used by Kleen Air Inc. The system conforms to the six elements of ISO 14001 and was written over a two-month period by G. Warner and D. Dilbert, deputy air quality managers and the most respected employees of the facility.

Written by:	Name	Revision Number:
Approved by:	Signature:	Date:

EMS Manual	Kleen-Air Inc.	Environmental Policy

It is the policy of Kleen-Air to preserve the cleanliness of the 350 political bodies we serve—as well as that of our immediate neighbors—by practicing the latest incineration safety techniques, such as scrubbing, available. We regularly stay apprised of new technology and employ it at our facility, and our up-to-date safety practices are verified by the Environmental Protection Agency as well as by local and state regulatory bodies. Our commitment to pollution prevention is unquestioned and ongoing.

Responsibility for ensuring compliance with regulation and policy requirements falls to the air quality manager and the general manager. They appoint others to stay abreast of these requirements, to procure industry research, and to scout bids on safety equipment.

Within the framework of our constantly improving practices is a commitment to identify and fulfill environmental objectives and targets and identify and minimize environmental impacts and aspects. Our employees are aware of this policy, and it is available for public inspection.

Written by:	Name	Revision Number:
Approved by:	Signature:	Date:

EMS Manual	Kleen-Air Inc.	Definitions of Terms

Scrubber: A pollution control device that controls air emissions.

EMS Manual	Kleen-Air Inc.	Element 4.1 General

4.1.a Purpose
The purpose of this section is to indicate that Kleen-Air operates under an environmental management system (EMS) and that that EMS is documented in this manual.

4.1.b Scope
This EMS applies to every aspect of the operation of the Kleen-Air Inc. incinerator in Anytown, USA.

A Sample EMS Manual **151**

4.1.c References
ISO 14001 Element 4.1, General

4.1.d Definitions
None

Written by:	Name	Revision Number:
Approved by:	Signature:	Date:

EMS Manual	Kleen-Air Inc.	Element 4.2 **Environmental Policy**

See prior pages for environmental policy.

Written by:	Name	Revision Number:
Approved by:	Signature:	Date:

EMS Manual	Kleen-Air Inc.	Element 4.3 **Planning** Subelements 4.3.1, 4.3.2, 4.3.3, 4.3.4

4.3.a Purpose

The purpose of this section is to indicate that the planning element is fulfilled through the determination of environmental aspects, legal and other requirements, and environmental objectives and targets, and the establishment of environmental management programs.

4.3.b Scope
This element affects the air quality department.

4.3.c References
ISO 14001 Element 4.3 and all subelements
Kleen-Air Environmental Aspects Register, Legal Requirements Journal, List of Objectives and Targets, and Document KA-V-1, Environmental Management Program Schedule

| Written by: | Name | Revision Number: |
| Approved by: | Signature: | Date: |

| EMS Manual | Kleen-Air Inc. | Element 4.4
Implementation and Operation |

4.4.a Purpose
The purpose of this section is to establish that management defines the structure and responsibilities of the environmental management system, human resources ensures and reinforces training and competence, public relations handles all internal and external communication regarding the EMS, the air quality department documents the EMS and controls all documents and all daily environmental operations, and the safety department has drawn up a safety response plan in case of emergency.

4.4.b Scope
This element applies to management and to the human resources, public relations, air quality, and safety departments.

4.4.c References
ISO 14001 Element 4.4 and all subelements

Written by:	Name	Revision Number:
Approved by:	Signature:	Date:

| EMS Manual | Kleen-Air Inc. | Element 4.5
Checking/Corrective Action |

4.5.a Purpose
The purpose of this section is to affirm that the air quality department monitors and measures the EMS for significant environmental impacts on a regular basis, defines procedures for detecting nonconformities and instituting corrective action, maintains records of all findings, and selects personnel to conduct internal EMS audits.

4.5.b Scope
This element affects the air quality department.

4.5.c References
ISO 14001 Element 4.5 and subelements 4.5.1, 4.5.2, 4.5.3, 4.5.4
KA-EA-1, Monitoring and measurement log

Written by:	Name	Revision Number:
Approved by:	Signature:	Date:

| EMS Manual | Kleen-Air Inc. | Element 4.6
Management Review |

4.6.a Purpose
The purpose of this section is to affirm that management reviews of the EMS take place on a regular basis, the aim being to continually improve the system.

4.6.b Scope
This element of ISO 14001 affects management—specifically, the general manager, air quality manager, chairman, and chief compliance officer.

4.6.c References
ISO 14001 Element 4.6 KA-MR-16, Management Review Agendas

4.6.d Definitions
None

Written by:	Name	Revision Number:
Approved by:	Signature:	Date:

EMS Manual	Kleen-Air Inc.	Circulation List

This manual is to be circulated only to the following personnel within Kleen-Air Inc.

- Chairman
- Chief compliance officer
- Air quality manager
- General manager
- Gerard Warner, Dennis Dilbert, authors

This concludes the Kleen Air Inc. EMS Manual.

Written by:	Name	Revision Number:
Approved by:	Signature:	Date:

You'll notice that we have made this EMS manual deliberately simple. Yours probably won't be this brief, but our example serves an important illustrative purpose: to demonstrate the necessary superficiality of the document.

Note that:

- We made reference to the elements without defining each subelement.
- We made reference to related documentation without including it.
- We used titles instead of names (in case of turnover, this saves you the trouble of making revisions).
- We haven't truly proven that these elements are in place. Only subordinate documentation can serve this purpose.

You are now equipped to build, from the top, the documentation pyramid for your facility's EMS. All other documentation should elaborate on what we have said here.

13
Market Forces and the Future of ISO 14000

Marketing and Practical Advantages of ISO 14001 Registration

There is no question that jumping on the ISO 14001 bandwagon can produce benefits for any company doing business internationally.

But registration to the standard is also a plus for businesses operating domestically.

In these environmentally conscious times, proving that you are paying attention to your organization's effect on the environment can set you head and shoulders above the competition—especially if your customers support ISO 14001.

If your customers not only support the standard but are registered to it and expect you to do the same, you have no choice. You may even find ISO 14001 registration popping up as a requirement in your customers' or suppliers' contracts.

Cutting costs is a high priority in any company, and this can be another benefit of an efficient ISO 14001 environmental management system. If, for example, avoiding high turnover—and higher training costs—is a priority for you, element 4.4.2 ("Training Awareness and Competence") of the standard may offer helpful guidance on how to maintain overall employee competence.

Lost or unintelligible paperwork can waste your time when you would rather focus on turning out your organization's product or service. The document control elements of ISO 14001 can be helpful here.

If you stay on top of regulatory requirements as specified by ISO 14001 element 4.3.2, and subsequently improve your environmental compliance, you may reduce government audits. Your registrar's stamp of approval may eventually be enough to convince regulatory auditors that their time would be better spent snooping around elsewhere. The U.S. Environmental Protection Agency (EPA) has indicated this possibility.

Perhaps most important, registering to ISO 14001 will help you control your organization's environmental aspects and protect the natural surroundings we all share.

Once these controls, such as those leading to pollution prevention, are permanently in place, your long-term regulatory compliance costs will drop, as will your exposure to legal liability. As an added bonus, if you are operating a company that is "environmentally friendly," your obligation for large-scale remedial cleanup projects may also diminish.

Your New Competitive Edge

> Business interests have now merged with environmental issues. Forward-looking companies are now building the environmental issue into their strategies. Forward-looking companies are now starting to make investments in research and development which will make their facilities and products more environmentally sound. Many of the research objectives necessary to achieve this goal will also lead to lower costs, higher quality, more marketable products, fewer liabilities, better employee morale and enhanced corporate reputation. Companies who do not include the environmental issue in their strategy risk losing their competitive position in the long run.
>
> —Robert P. Bringer, staff vice president of environmental engineering and pollution control for 3M, from a 1991 address at Yale University.[*]
>
> In a 1995 Arthur D. Little survey of 115 large North American businesses, 61% expected that meeting ISO 14000 would bring "potential competitive advantage," and 48% said that not meeting the standard could constitute a "potential nontariff trade barrier" because customers around the world are expected to require it.[†]

---

[*]Steven J. Bennett, Richard Freierman, and Stephen George, *Corporate Realities and Environmental Truths: Strategies for Leading Your Business in the Environmental Era*, Wiley, New York, 1993, p. 19.

[†]Ronald Begley, "ISO 14000: A Step Toward Industry Self-Regulation," *Environmental Science & Technology*, vol. 30, no. 7, 1996, p. 299.

Throughout this book we have referred to the competitive advantage of ISO 14000 registration. This advantage will become increasingly pronounced as the "outside world" becomes more aware of the standard and the virtues of internal environmental management.

This should happen in the United States around the end of 1998, said Joel Charm, leader of the Product Stewardship Center of Excellence at Allied Signal in Morristown, New Jersey, and a member of TC 207 (although it is certainly useful *now* as a tool for attracting potential customers and suppliers who have subscribed to the tenets of ISO 14000).

Parameters of Using ISO 14001 in Advertising

Presuming that you have read and used the advice in this book, and your facility has become ISO 14000–registered, you will now need to understand the parameters for your usage of the standard in advertising. In this chapter, we will also relay some common sense advertising advice we found elsewhere.

One of the most direct ways companies choose to advertise is by slapping colorful, snazzy, exclamatory labels directly on their products. When the country was swept by the low-cholesterol craze, companies began labeling their products in a way intended to reassure their customers that their foods were safe to eat. Go to any grocery store. Even sweet, fattening, long-time-favorite junk foods like cookies, baking mixes, and even some candy bars contain labeling that tells you that the product is now "low cholesterol" or "contains 33% less fat" than other brands. Slick television commercials serve to bolster these claims. Upon close inspection, however, a smart shopper finds that the product is loaded with sugar and saturated oils—other dietary ills that act to raise cholesterol.

Less harmful but equally disturbing, many products' labels claim that the product is "new and improved" or that the container is filled with "18% more free" product. How does one prove this? And why is it important to buy a car that goes from "zero to 60" in the least amount of time? These claims are, to the savvy consumer, rather meaningless.

Unfortunately, in much advertising, the use of exaggerated, false, and misleading claims are the norm. Companies that want to indicate their commitment to the environment use buzzwords like "green" and "environmentally friendly." One of the areas in which this is most prevalent is the converting of products (like household cleaning products) which are traditionally dispensed from pressurized metal spray cans whose gases are harmful to the ozone layer, to plastic trigger-pump bottles.

Presidential candidates in the modern age claim to be "environmentalists" if they think it will boost their stock with environmentally oriented voters. What these people do to prove that commitment, if anything, is another story. (Ironically, in 1996, President Bill Clinton designated a large area of land in Utah, near the Grand Canyon, as off limits to development. While this pleased naturalists, it upset nearby residents who depend upon the land's riches for economic survival—proving once again that environmental preservation can clash with prosperity.)

The business representatives responsible for the formulation of the ISO 14000 series, however, wrote *ISO 14020—Environmental Labeling* to set specifications for the use of environmental claims in one type of advertising: labeling. Let's take a general look at those provisions, and then at the advice of other environmental experts.

Labels are categorized into three types under ISO 14020:

Type 1: Third-party certified environmental labeling

Type 2: Informative self-declaration claims

Type 3: Specific product information that has been independently verified

No matter which category is used, and labels of this nature take many written and graphic forms, all claims must be substantiated. A tuna manufacturer, for example, cannot advertise that its product is "dolphin-friendly" simply to capitalize on naturalists' opposition to dolphins being trapped and tangled in tuna-catching nets. If that manufacturer avoids the capture of dolphins, the label must indicate something specific to that effect. A more appropriate label would say something like this: "Since 1994, XYZ Tuna has modified its netting procedures such that no dolphins have been trapped in our apparatus. This information has been verified by the U.S. Fish and Wildlife Service, and its research is available for public inspection (see item *h*, below)."

Try fitting that on a tuna fish can next to the picture of the mermaid. It runs counter to everything advertising executives learn about catching the eye of the buying public.

As you immerse yourself in a study of the standard, and of ISO 14020 in particular, you will see that labeling guidelines and requirements are actually delineated in the following subordinate documents:

ISO 14020: Environmental Labeling—Basic Principles of All Environmental Labeling. Provides general guidance to help organizations develop specific environmental claims. ISO 14020 streamlines the guidelines of the

Federal Trade Commission (FTC) and can be used as a tool by purchasers for selecting products and services.

ISO 14021: Environmental Labeling—Terms and Definitions. This document eliminates confusion by providing definitions for the various labeling technologies in use.

ISO 14022: Environmental Labeling—Self-Declaration Environmental Claims—Symbols. Addresses the use of symbols in self-declaration environmental claims and provides general principles for using environmental symbols. This document is also aligned with FTC guidelines.

ISO 14023: Environmental Labeling—Self-Declaration Environmental Claims—Testing and Verification Methodologies. Provides general principles for carrying out testing and verification methodologies to substantiate environmental claims. Such methodologies must be reproducible, repeatable, and scientifically sound.

ISO 14024: Environmental Labeling—Guiding Principles, Practices, and Criteria for Multiple Criteria-Based Practitioner Programs (Type 1)—Guide for Certification Procedures. Provides criteria for practitioner programs such as Green Seal or Blue Angel. This standard is intended to help practitioners evaluate products and award labels to companies.

Simply put, the ISO 14020 series seeks to make the use of environmental claims appropriate by

- Providing accurate, verifiable, and nondeceptive environmental claims
- Minimizing unwarranted claims
- Reducing marketplace confusion
- Enabling purchasers to make informed choices
- Reducing restrictions on and barriers to international trade
- Increasing the potential for market forces to stimulate environmental improvements in products, processes, and services

In order to fulfill these requirements, any and all environmental claims that a company attaches to its product or service (and remember that these rules are adaptable to all manner of commerce) must:

1. Be accurate and nondeceptive
2. Be substantiated and verifiable
3. Be relevant to that particular product or service, and be used only in an appropriate context or setting

4. Be specific and clear as to the particular attribute to which the claim relates
5. Be unlikely to result in misinterpretation
6. Be meaningful in relation to the overall environmental impact of the product or service during its life cycle
7. Be presented in a manner that clearly indicates that the environmental claim and explanatory statement should be read together
8. Not be presented in a manner which implies that the claim has been endorsed or certified by an independent, third-party organization when it has not been

Some examples of claims that are disallowed under ISO 14020 are "environmentally friendly," "ozone friendly," "nature's friend," "earth friendly," "nonpolluting," and "green."

Even before the development of ISO 14000, environmental well-being was a topic of debate. At least one author has written that the environment replaced nuclear testing at the top of the list of special interest groups' concerns as early as the 1960s (Stroup, p. 195).* In *Corporate Realities and Environmental Truths*, Steven J. Bennett, Richard Freierman, and Stephen George wrote that companies must be prepared to back up their claims if they wish to capitalize on consumer preferences for products that are manufactured and distributed with the environment in mind. They write that a backlash occurred when consumers became distrustful of marketing "fiascoes" that made use of ambiguous words and catchphrases (pp. 92, 93).

But a 1992 poll conducted by *Advertising Age* magazine indicated that 63 percent (p. 93) of consumers were still interested in environmentally sound products (Bennett, Freierman, and George didn't indicate the size, scope, or margin of error of the *Advertising Age* study).

Advertising with an environmental twist (so-called eco-marketing) is hardly similar to claiming that your dishwasher detergent is new and improved or contains 30 percent more product. Your campaign is likely to attract at least routine scrutiny from both law enforcement agencies (which also judge the superficial claims we discussed earlier) and environmental watchdog groups. You may be opening a Pandora's box by exposing your ads to a *variety* of groups whose research will vary widely from yours and from one another's. You may be starting down a slippery slope of contro-

---

*Richard Stroup, in Walter E. Block, ed., *Economy and the Environment: A Reconciliation*, Fraser Institute, 1990.

versy, the authors warn, by inviting doubt about your intentions and the positive environmental attributes of your products (p. 94).

So your approach must not be one of "marketing as usual" (p. 94). Quite the contrary, you should be prepared to educate yourself, compile research and statistics, and even consult knowledgeable experts who can predict the emotional impact of the issues you will be raising. Bennett, Freierman, and George, even suggesting going so far as to assemble an expert panel to review your ad campaigns before they are exposed to the outside world.

Be sensitive to your audience's environmental conscientiousness. The authors write:

> An important prerequisite to eco-marketing is to understand that people who are willing to make lifestyle changes, even simple ones, to improve the environment in some small way feel a sense of personal satisfaction and commitment. They feel that they're doing their part and the companies they buy from should show similar efforts in reaching the common goal of a cleaner and safer environment. Consequently, when they read about corporations exploiting the environment through deceptive or misleading advertising, they become cynical and even incensed (pp. 94–95).

Bennett, Freierman, and George's 10 laws for eco-marketing (pp. 96–102) are remarkably similar, if broadly stated, to those that appear in ISO 14020:

1. *Avoid nebulous slogans.*
2. *Don't invent a "tradition of environmental excellence."* It doesn't matter that your company's founders lived primitively in a log cabin and hunted and gathered. What are you doing to prove your environmental concern today?
3. *Watch your language.* As we said above, the words you choose can provoke suspicion. Terms like "biodegradable" have a specific meaning. Don't bandy them about cavalierly.
4. *Give the truth, the whole truth, and nothing but the truth.* Any warm-fuzzy claim you make or advertising slogan you use that has to be qualified is as good as a lie.
5. *Pick your numbers carefully.* The fact that there are hundreds of thousands of doctors in this country renders the "three out of four doctors" advertising cliché meaningless. Talking to four doctors and getting agreement from three hardly represents overwhelming approval or serves as valid scientific research.

6. *Build credibility.* Do this by inviting external verification of your facts.
7. *Be a partner, not a teacher.* Don't pontificate as if only you have all the facts. The general public, and your stakeholders in particular, probably have some knowledge about these issues.
8. *Show relevance.* The fact that you have contributed $5,000 to Greenpeace doesn't make your brand of motor oil a better product.
9. *Don't tout the trivial.* The authors of *Corporate Realities* call this the most scoffed-at law of eco-marketing and have quite a bit to say about it. They cite the example of an airline that bragged after Earth Day 1990 that it recycles its beverage cans. While this is noble, the airline failed to acknowledge the much larger problem of atmospheric pollution caused by jet aircraft.

    In attempting to sift the trivial from the nontrivial, ask yourself these questions (pp. 100–102):

    - Does the action about which you are boasting solve a widespread, well-known environmental problem?
    - Does your product or service go above and beyond the call of environmental compliance?
    - Are your company's environmental accomplishments substantial and far-reaching? Is a $10,000 donation to the restoration of Prince William Sound's ecosystem really substantial, considering that the true cost of the effort runs into the billions of dollars?

10. *Make eco-marketing part of a sustained effort.* Is your eco-marketing part of a sustained effort, or is it an isolated flash-in-the-pan novelty? Is it motivated purely by environmental goals, or are there other factors, such as the desire for a community service award plaque for your outer office wall? In other words, are you being altruistic or opportunistic?

Even if you are inclined to implement environmental management that is only *based upon* the tenets of ISO 14000, if you want to promote your new-found environmental-mindedness, these rules must still apply. It is the only way to ensure the integrity of the concepts of sound environmental management.

If you earnestly seek better environmental performance, an improved image for your organization, reduced costs and regulatory burdens, a tremendous competitive edge, and a decreased likelihood of litigation, there is no better way to achieve these goals than to start planning your ISO 14001 education and registration today.

14
Conclusion

ISO 14000 and Social Responsibility

The ISO 14000 series was written with an eye toward consistent environmental management leading to improved environmental performance, which has inherent *commercial* benefits: reduced compliance costs, improved competitive positioning, fewer customer and regulatory audits, better public image, and increased profitability. While social responsibility may not have been part of the standard's blueprint (although it's worth noting that, according to Allied Signal's Joel Charm, the pollution prevention portions were suggested by representatives of the Environmental Protection Agency), a facility registered to ISO 14001 may be rewarded intrinsically by knowing that it has accomplished that hurdle as well.

Indeed, as early as 1960 there was a recognition in some quarters that corporations as a whole must balance their profit margin with a certain amount of attention to the outside world from which they draw that profit. This philosophy has led to the doctrine of corporate social responsibility (CSR), which according to W. C. Frederick requires that businesses

> oversee the operation of an economic system that fulfills the expectations of the public. And this means in turn that the...means of production should be employed in such a way that production and distribution would enhance total socio-economic welfare. Social

responsibility [is] a willingness to see that resources are utilized for broad social ends and not simply the narrowly circumscribed interest of private persons and firms.*

There is also, of course, a famous precept that businesses as private institutions do in fact have a responsibility, but that that responsibility extends only to increasing the profits of their owners and shareholders. It has been our purpose here, even if indirectly, to encourage and facilitate your registration to ISO 14001 and to illustrate that the standard allows you to increase profits by doing so. As an added bonus, we have also sought to understand the environmental "movement," its continued evolution, and its effect on today's business climate.

It is ironic that the release of the ISO 14000 series coincided with a "down" period in blind and unbridled enthusiasm for environmental issues—especially considering that in no way was this deliberate or planned. The literature suggests that in the 1990s, society's focus has shifted away from altruistic concerns about the outside world to self-preservation in the face of economic uncertainty. Still, by subscribing to the tenets of the standard, you are not declaring yourself to be "green," nor will you be labeled as such. It's simply a decision that makes good business sense and sharpens your competitive edge.

While we're on the subject, as we have pointed out, the standard explicitly discourages such ambiguous publicity about your newfound environmental conscientiousness as calling yourself a "green" or "Earth-friendly" facility. By registering to ISO 14001, you are letting the world know that you are prepared to document and validate your improved environmental performance.

And you needn't be a corporate giant like General Motors or Dow Chemical, or any other company with multiple environmental impacts, to find validity in ISO 14000 or to see its potential benefits (General Motors met to consider ISO 14001 implementation but at this writing was undecided). It is a standard that is highly adaptable to any size or type of facility, and to any corporate aim. Whether you are looking to build a larger, international client base or to advertise to the world at large that you are now an environmentally responsible corporate neighbor, ISO 14000 can work for you.

---

*W. C. Frederick, "Implications of Environmental Management Strategy for Manufacturing Performance," University of North Carolina, Chapel Hill, 1995, p. 16.

Recap of Implementation Steps

By following a few simple steps, you can begin to put the standard in your corner. Let's capsulize these.

Select an employee to research and learn about ISO 14000. As this book appears, there is bound to be a plethora of available training, both overview-type seminars and intensive week-long courses, as well as a foundation of literature focusing on ISO 14000 and related issues. If you are truly serious about the standard, give at least one diligent employee some time away from his or her regular responsibilities to focus on what the standard means and whether it is suitable at your facility at this point in time.

Select an implementation team to receive training with eventual registration in mind. The first step was merely an informational outing; now that you have decided to put ISO 14000 to work in your facility (we reiterate that ISO 14001 registration is by facility, not by firm), put together a highly competent team of dedicated individuals to attend implementation and auditor training so that they can return to work ready to plot a timetable leading to registration.

In a larger, highly structured facility, this team is likely to comprise your entire management (including environmental management) staff. You need not include front-line environmental performance "operators," who will carry out the strategies of your ISO 14001 EMS. While this sounds exclusionary and elitist, and goes against the principles of total quality environmental management (TQEM), it is perfectly in line with the fact that ISO 14001 is an environmental *management* standard, to be implemented by management.

Do a gap analysis. One of the first things your implementation, or steering, team will learn in its training is that you must determine how far your facility currently is from operating according to ISO 14001 and what it will take to fill those "gaps." If you are a facility that makes environmental impacts, you probably have some sort of environmental management system in place already. Rounding out that system with the requirements of the ISO 14001 specification standard is what implementation is all about.

Write necessary documentation in accordance with the standard's documentation requirements for document control, etc. Another way of explaining the purpose of ISO 14000 is: "Say what you do. Do what you say. Prove it." In other words, it isn't enough to have an environmental management system that runs smoothly. Under ISO 14000, you must document the system as a way of proving that it exists (say what you do).

Then you must actually operate according to the dictates of that documentation. It isn't sufficient to say that you will have periodic management reviews, as required by element 4.6 of the standard. You must actually hold them (do what you say).

By virtue of EMS records, the last of the documentation tiers, you *prove* that such management reviews were held. You *prove* that you are monitoring and measuring your EMS by keeping records of those measurements (prove it).

Operate and fine-tune your EMS over a period of three or four months. Do this in order to accomplish the following: Generate the above-mentioned records, which will serve as solid proof of an EMS, and get used to the new system and its requirements of vigilance and continual improvement.

Learn how to conduct and then carry out internal audits. Do this in order to determine the effectiveness of your EMS.

Contract with a registrar. You will have a lengthy relationship with this registrar through the registration audit and periodic surveillance audits.

You will keep an eye on continual improvement, a subject no book on ISO 14000 could possibly overemphasize. Continual improvement is the philosophical underpinning of the standard. It means that you must constantly determine the effectiveness of your EMS and make necessary changes. You will certainly have your eye on continual improvement between surveillance audits, when your registrar drops in to make sure you are still worthy of ISO 14001 registration status.

Will you do it all alone, as we have suggested above, or will you use a consultant to keep you on track? As we have pointed out, consultants are industry insiders who can point you in directions you never knew existed. How you choose a consultant depends upon your needs and how far you are willing to go to find a competent, qualified one. As the standard becomes more universally accepted, more consultants with environmental expertise are likely to educate themselves in its principles and make their services available.

There will also be fresh-faced "experts" who are just starting to get their grounding in environmental issues, let alone ISO 14000 and its specification standard, ISO 14001. These people may not be your best choices.

Be patient. The world at large has not yet caught onto ISO 14000 (although it is widely recognized in Europe and the EU), and the advertising benefits of being registered may not be available immediately. When you do begin to publicize your status, you will find that the primary benefit will be enlarging your supplier and/or customer base.

If you find that you are floating in a sea of unregistered suppliers, you may be able to convince them to also undertake registration and hence create a supplier chain. If you cannot, seek out those suppliers who at least adhere to the principles of sound environmental management. You'll be free to set those criteria yourself.

Scholarly Research in Environmental Management

The term *environmental management* wasn't invented with the creation of ISO 14000. It is a subject which has received years of scholarly attention and research, and this research hasn't always supported the notion that companies can predictably improve their economic well-being by paying attention to their effects on the outside world.

"The study determines that superior environmental performance is related to superior economic performance only for certain types of environmental performance and, more particularly, certain types of environmental performance within certain industry sectors. On the other hand, in some cases, poorer environmental performance seems to be economically rewarded," writes Scott David Johnson in his 1995 doctoral dissertation (pp. xxi–xxii).[*]

While Johnson acknowledges that poor corporate performance is linked to higher costs and, conversely, that exceptional environmental performance breeds lower costs and greater efficiency (p. 1), he suggests that those higher costs can be balanced out by a better economic showing.

> For example, although oil spills negatively impact economic performance, poorer environmental performance (e.g., higher emissions) is actually consistent with better economic performance in the case of total toxic chemical emissions. These results suggest that the present

---

[*]Scott David Johnson, "An Analysis of the Relationship between Corporate Environmental and Economic Performance at the level of the Firm," doctoral dissertation, University of California—Irvine, 1995.

U.S. environmental regulatory system is not effective in creating consistent economic incentives for corporations to allocate resources toward the improvement of their environmental performance, particularly with respect to toxic chemical emissions (p. 187).

Johnson adds that the aerospace/transportation equipment, apparel/textile, jewelry/metal product, motor vehicle, and pharmaceutical industries are most likely to suffer economically for high emissions.

Johnson writes that managements have traditionally placed environmental performance on the back burner, preferring to focus on growing their businesses, but today realize its importance for several reasons (pp. 3–4):

- Both the explicit and the public relations costs of environmental accidents are increasing.
- Costs for natural resource damages above and beyond actual cleanup costs are being levied against companies in more and more court settlements.
- The expected future costs of environmental liabilities (e.g., Superfund site cleanups) are creating a burden on corporate balance sheets as a result of Securities and Exchange Commission (SEC) disclosure requirements.
- The costs associated with fines, penalties, and judgments and associated legal defenses for noncompliance with environmental laws and regulations have been increasing, and recently sanctions have extended beyond civil penalties to criminal charges affecting employees at all levels of the organization.
- Disposal costs for industrial wastes are increasing significantly as treatment standards at disposal sites are raised.
- Firms with poor environmental records have greater difficulty in obtaining financing and insurance (e.g., Bank of America claims to consider environmental performance in loan decisions).
- Public demand for environmentally superior products and firms is increasing and, likewise, public disdain for firms that are perceived as polluters is increasing.

The government itself, as of 1986, requires that facilities report some of their emissions to the EPA (p. 50).

Despite this new environmental conscientiousness, and the fact that recent surveys show managers placing the environment high on their list of priorities (p. 6), Johnson writes, remarkably, that "the belief that environmental protection is simply an added cost to be avoided wherever possible is still alive and well at all operational levels of American corporations today" (p. 4).

Johnson set out to answer the question of whether companies that demonstrate environmental excellence are generally more successful. The only way this can become an established pattern, he says, is for companies that fare better environmentally to also benefit economically. Meaningful cost-benefit calculations for improved environmental performance, however, are elusive because so many of the benefits of which Johnson speaks (e.g., aesthetics, preservation of species and biodiversity, lower health-care costs, protection of groundwater) are hard to quantify and are external to the current money-oriented system in which companies operate. Hence, he suggests, a new kind of economy that values environmental protection must be created (p. 5).

This will be difficult if Johnson's key observations continue to be true. These are (p. 197):

- Legal chemical emissions have an environmental impact two factors of magnitude *greater* (emphasis added) than the impact of chemical spills.
- The average actual dollar value of the regulatory fines levied upon the average *Fortune* 500 firm under the five key federal environmental statutes is negligible.
- Approximately half the toxic chemical emissions by major corporations are emitted to media with the potential for direct human and ecological contact.
- The greatest volume of emissions, as a function of sales, were due to the following major industries (listed from highest to lowest levels of emission): chemicals, mining/crude oil/petroleum, jewelry/metals/metal products, pharmaceuticals, building materials/forest products, furniture.
- Firms with higher emissions of toxic chemicals do have a statistically higher level of regulatory violations. However, the association is relatively weak.

Johnson concludes by saying that the present system of environmental sanctions in the United States is not sufficient to give corporations an economic incentive to improve their environmental performance overall.

Robert David Klassen, in his dissertation titled "The Implications of

Environmental Management Strategy for Manufacturing Performance,"* concluded that firms using *adaptive,* or proactive, environmental management technologies fared better in terms of manufacturing performance (presumably leading to higher profits) than firms which employed *conventional,* or end-of-the-line, inspection-oriented processes.

Adaptive technologies involve product and methodological changes to existing steps in the manufacturing process. Some examples of these, with their associated risks, are charted in Fig. 14-1, an illustration that originally appeared in Klassen's dissertation.

Conversely, facilities employing only the conventional, reactive strategies did not do as well. The downside of adaptive technologies, however, which more closely match the objectives of ISO 14000, was lower quality (p. iv).

Obstacles to Investment in Preventive Technologies

While preventive technologies are highly touted as a cost-effective alternative, several types of obstacles to investment in them have been highlighted (p. 3):

> *Economic obstacles.* Uncertainty over long-term investments in technological change makes them more difficult to plan and commit to than simply adding processes at the end of the line.
>
> *Information obstacles.* Management may not be aware of the adaptive options; in fact, Klassen's study revealed that management attitudes and commitment are the most important factors in determining environmental performance (p. 27).
>
> *Management obstacles.* New technologies are seen as contrary to bottom-line business objectives.

In "Enterprise, Value, Environment: The Economics of Corporate Responses to Environmentalism,"† Blair Witten Sandler writes that companies have sought to stymie environmental legislation (p. 2). Sandler argues that this philosophy can lead to environmental destruction (p. vi).

---

*Robert David Klassen, "The Implications of Environmental Management Strategy for Manufacturing Performance," Doctoral dissertation of North Carolina, Chapel Hill, 1995.

†Blair Witten Sandler, "Enterprise, Value, Environment: The Economics of Corporate Responses to Environmentalism," doctoral dissertation, University of Massachusetts, Amherst, 1995.

Category of Technology	Contribution to Sustainable Development	Target of Innovation	Environmental Risk	Production Risk	Examples
Remediation	Limited	Off-line process	Limited	None	Removing leaking underground storage tanks
End of Pipe Pollution Controls	Limited	Process related	Limited	Limited	Electrostatic precipitator for air discharge stack
Managements Systems	Mixed	Management/Human	Moderate	Moderate	Plant-level environmental audit
Production Adaptation	Significant	Product related	Significant	Significant	Replacing an organic based paint with water based paint
Process Adaptation	Significant	Process related	Significant	Significant	Recycling process water

Figure 14-1. In his 1995 doctoral dissertation for the University of North Carolina, Chapel Hill, Robert Klassen listed examples of preventive environmental technology and the risks associated with each.

This brings up another point: Not only has management traditionally seen environmental issues as obscure and ancillary to "real" business concerns, Klassen writes, but such individuals have often regarded the environment as strictly a legal and regulatory compliance issue—another mindset that ISO 14000 seeks to change.

Some regulatory measures, ironically, may bring environmental management into a holistic and competitive focus. Klassen writes that "an operations strategy that integrates environmental management can establish or alter barriers to entry for other industrial competitors" (p. 14). And, because government agencies are lowering the acceptable levels of some emissions, firms have an opportunity to lead their industries by acquiring new technologies, developing in-house management and control systems, and influencing the adoption of environmental products, processes, and management systems (p. 15). Another model, he writes, similarly suggests that "manufacturers who demonstrate efforts to minimize the negative environmental impacts of their products and processes through using better design, recycling post-consumer waste, and establishing environmental management systems are poised to expand their markets or displace competitors that fail to promote strong environmental performance" (p. 30).

In terms of costs, Klassen writes, "Firms that invest heavily in environmental management systems and safeguards can potentially avoid future environmental spills, crises and liabilities. Cost resulting from waste materials and inefficient processes are also minimized" (p. 30).

Because of the newness of environmental management techniques to companies used to operating reactively, those companies' initial "naivete and lack of sophistication" require sticking with reactive strategies to some extent. Later, however, with education, training, and orientation to standards like ISO 14000, companies and facilities can become more responsive to external stakeholders. The learning curve looks like that in Fig. 14-2, again, from Klassen.

Klassen adds that most firms are likely to fall on the diagonal as they move from a reactive strategy to becoming more innovative.

Klassen, whose research was conducted within the furniture industry, concluded that plants whose management personnel favor proactive, adaptive environmental technologies may have better manufacturing performance than those who follow reactive, conventional technologies.

He said, however, that the connection between adaptive technologies and improved environmental performance was "not as clear-cut" (p. 173), adding that vast differences were found between subindustries

Conclusion

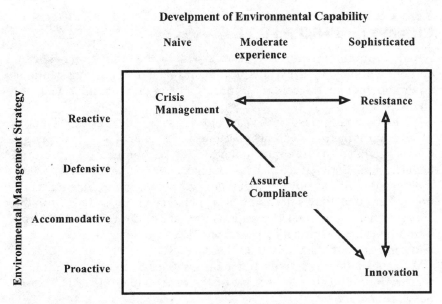

Figure 14-2. Robert Klassen demonstrated that a facility becomes more environmentally proactive as it becomes more familiar with sound environmental management techniques.

(home furniture vs. office furniture, etc.), with the clearest correlation occurring among office furniture manufacturers. He writes that, with its use of raw materials, highly toxic finishes, and other substances, the furniture industry makes a large environmental impact.

Why, then, with this and other highly technical research drawing only shaky conclusions about the virtues of proactive environmental management, do we advocate its use? Why not simply encourage corporations and their subsidiary facilities to pay more attention to their *current* environmental programs?

This is because ISO 14000 is the only uniform environmental management system with specific requirements that are applicable across the board to companies of all shapes, sizes, and functions. ISO 14000's underlying consistency means that classes can be offered and advice of a one-size-fits-all nature can be given and adapted to any facility with allowances for the extent of the environmental management system a facility is currently using.

The Changing Economic Climate and ISO 14001

A new corporate climate, now presided over by the "baby-boomer" generation, will play an especially prominent role in driving conscientious environmental management, according to environmental consultant Denise Seipke.

"They're getting close to 50 years old. These people grew up at a time when environmental awareness became an issue, and as a result people of this age have an environmental ethic. It's part of their own personal belief system. So what happens is, when a company goes to ISO 14000, they are in essence putting together a corporate environmental ethic policy. And that allows the employees' personal ethics to mesh with the ethics of the company. Corporations are made up of individuals and individuals have ethics; it's that simple," she said. "I think there are certain companies that already hold these ethics."

And ISO 14000 will give companies a personal investment through registration, replacing their view of the environment as a liability of heavy regulation.

"For most companies, regulation has been a constant source of irritation. It's an ongoing process that's never-ending," Seipke said. "What happens is, as you show companies a broader picture, as you show them how to implement ISO 14000 and show them the benefits that can be derived from total environmental assessment rather than just regulatory compliance, you see a change of attitude. The whole point of this is to be able to change your role.

"If you work closely with the entire community and you accept your impacts and look at how you're going to change those impacts, you'll end up with PR you couldn't buy with all the money in the world. People understand the difference between those companies that don't care and those that do," she said.

She acknowledges, however, that attitudes must change to the point where ISO 14000 is understood and accepted.

"Attitude is half the battle. If you talk to children about recycling, it becomes an internalized ethic. What ISO 14000 does is take the environment to a new dimension where the intent is to be a good corporate citizen. Certain utility companies, for example, have landscaped their grounds for wildlife. So have certain manufacturing companies. A lot of companies provide financial support to environmental education in the public sector through informal education such as the Scouts, and then again through school programs where they donate (money) for software, laboratory development, that sort of thing. It's not enough just to meet the regulations; we're concerned with total impact.

"It isn't a matter of preserving this pristine environment, but about operating in a conservative manner that allows you to have your cake and eat it too," Seipke said.

The reason ISO 14000 is going to succeed, she said, is that people and companies do care—not only in a dollars-and-cents way, but also with regard to their role in the community. But the so-called baby boomers, who have become more pragmatic and shifted their focus from work to personal well-being, will be the driving force behind the standard's adoption.

"People today, in their 40s, all grew up with the mindset that if you worked hard on the line, eventually you would have a management job. When all these baby boomers got up to this age, there were too many people, and the corporate mindset had changed so that there were no more middle management jobs.

"At that point, you have to have a reason for living. If I'm not going to reach these financial goals I set, I have to rethink my priorities. Quality of life has become more of an issue to people than it was earlier, because we have nothing else to focus on. We focus on why we're living and how much we're enjoying it. That means family is more important, leisure time is more important, and leisure time is spent interacting with the environment—walking, birdwatching, depending on what people like to do. So their environmental ethics are affected that way.

"So you're going to see a merger between individual ethics and corporate ethics. The bottom line is, environmental impact minimization is where they're *both* headed. And this is the first time you've seen this in the industrial world. We've never had this before."

Index

Accreditation
  agencies for auditors, 22–27
  Dutch Council for Accreditation (RvA), 22, 63–64
  Environmental Auditors Registration Association (EARA), 22–26
  of registrars, 63–64
  Registrar Accreditation Board (RAB), 63
  United Kingdom Accreditation Service (UKAS), 18, 64
Advertising
  rules for eco-marketing, 163–164
  using ISO 14001 in, 159–164
Air emissions, 80
Akzo Nobel Chemical, 4–5
American National Standards Institute (ANSI), 63
ANSI, 63
Associate environmental auditor, 23–25
Audit reports, 19–20
  nonconformity report (NCR), 12–13, 19–20
  registration, 20
Auditors
  accrediting agencies for, 22–27 (*see also* Accreditation)
  advice for internal, 11–12
  associate environmental, 23–25
  collecting objective evidence, 10
  environmental, 25
  internal, 35–37, 109–112
  methods for asking questions, 11
  principal environmental, 25–26
  qualifications
    for EMS, 16–18
    for internal, 11–12, 13–14
  team of, 10
  training for internal, 13–14
Audits, 9–39
  benefits of, 37–38
  conclusions arrived at during, 12–14
  definition/description of, 10
  desktop, 15
  EMS, 27–32, 56–57
  environmental, 37
  first-party internal, 9–10
  of compliance with company environmental policies, 30–31
  of compliance with environmental regulations, 30
  of environmental management systems, 31
  of environmental performance of other parties, 32
  of financial accounting for environmental liabilities, 32
  of pollution-prevention and waste-minimization programs, 32
  of risks and liabilities of property transfers, 31
  of treatment, storage, and disposal of hazardous wastes, 31
  preassessment, 15
  preregistration, 15
  reasons for conducting, 30
  registrars and, 60–62
  reporting guidelines, 38–39
  second-party, 20–21, 32, 37, 110
  social, 28
  surveillance (*see* Surveillance audits)
  third-party registration (*see* Registration audits)
  transactional, 31

Begley, Ronald, 158
Bennett, Steven J., 117, 118, 158, 162, 163

Biosphere, protecting the, 33
Bird, Warren, 107
Bosso, Christopher J., 97
Bringer, Robert P., 158
Brown, Carol, 29, 74, 76, 77, 78
Burford, Anne, 96

Campbell, Sharon Nelms, 28
Carpenter, George, 118
CERES, 33–35
Certification (*see* Registration)
Charm, Joel, 159, 165
Chernobyl nuclear accident, 97
Clinton, Bill, 98
Clinton administration, 29
Coalition for Environmentally
  Responsible Economics (CERES),
  33–35
Communications, 52, 129
Conformity, 12
Conservation, 34, 80
Consultants, 124–126
Contamination, 81
Corporate social responsibility (CSR),
  165–166
CSR, 165–166

Deming, W. Edwards, 118
Desktop audit, 15
Documentation
  EMS manual, procedures and record-
    keeping, 129–132
  environmental policy, 126–127, 129
  preparing, 129–132
  pyramid of ISO 14001, 130
  recordkeeping, 56, 131–132, 168
Donaldson, John, 63
Dutch Council for Accreditation (RvA),
  22, 26–27, 63–64

EARA, 22–26
Eco-Management Audit Scheme (EMAS),
  2, 18, 74
EMAS, 2, 18, 74
Emergency procedures, 54–55
Emergency-response planning, 3
EMP, 48–49, 127–128
EMS (*see* Environmental Management
  System)

EMS manual, 129–130, 143–155
  comparisons with subordinate docu-
    ments, 146
  example of an, 147–155
  footers for, 147
  headers for, 146
  importance and necessity of, 143
  ISO 9000 example of, 144
  numbering systems for, 147
  planning, 145–147
  steps in writing an, 144–145
Energy conservation, 34, 80
Environmental activism, 98–101
Environmental Auditors Registration
  Association (EARA), 22–26
Environmental committees, forming,
  119–120
Environmental hazards
  Chernobyl nuclear accident, 97
  Exxon Valdez oil spill, 1–2, 5, 97
Environmental labeling, 160–163
Environmental management, scholarly
  research in, 169–172
Environmental Management Program
  (EMP), 48–49, 127–128
Environmental Management System
  (EMS), 46
  audits of, 31, 56–57
  benefits of, 4–5
  documentation for, 52–53
  management representative's responsi-
    bilities, 50–51
  manual for (*see* EMS manual)
  origin of EMS auditing, 27–32
  procedures and practices of, 131
  recordkeeping, 131–132
Environmental policy, 129
  preparing an, 126–127
Environmental Protection Agency (EPA),
  28, 29, 89–95, 158
  accountability of, 93
  House and Senate committees and, 92
  ISO 14000 and, 94
  role of, 92–95
Environmental restoration, 34
Environmental risks, reducting, 34
Environmental standards, influences on, 2
EPA (*see* Environmental Protection
  Agency)
Epstein, Marc J., 134
European Union (EU), 2
Exxon Valdez, 1–2, 5, 97

Index

Fortlage, Paul, 63
Frederick, W.C., 165, 166
Freierman, Richard, 117, 162, 163

GATT treaty, 2
GEMI, 117
General Agreement on Tariffs and Trade (GATT), 2
George, Stephen, 117, 162, 163
Global Environmental Management Initiative (GEMI), 117
Grover, Ralph, 84, 104, 129, 140

Hess, Joe, 74, 76, 77, 79
Higby, Kay, 5
Hoy, Patrick, 75
Hunt, David, 45

IATCA, 65–67
Internal auditors, 35–37
    training program for, 109–112
Internal audits, 9–10
International Auditor and Training Certification Association (IATCA), 65–67
ISO 9000, 5, 137–142
    history of, 137–138
    quality manual for, 144
    relationship between ISO 14000 and, 139–142
ISO 9001, elements of, 138
ISO 14000, 2
    relationship between ISO 9000 and, 139–142
    social responsibility and, 165–166
ISO 14001
    Akzo Nobel Chemical's registration, 4–5
    benefits of, 6–7
    changing economic climate and, 176–177
    Coast Guard guidelines and, 3–4
    companies seeking registration to, 6
    Element 4.1 General, 42
    Element 4.2 Environmental Policy, 42–43
    Element 4.3 Planning, 43–49
    Element 4.3.1 Environmental Aspects, 43–44
    Element 4.3.2 Legal and Other Requirements, 44–48
    Element 4.3.3 Objectives and Targets, 48

ISO 14001 (*Cont.*):
    Element 4.3.4 Environmental Management Program, 3, 48–49
    Element 4.4 Implementation and Operation, 49–55
    Element 4.4.1 Structure and Responsibility, 4, 49–51
    Element 4.4.2 Training, Awareness and Competence, 3, 51–52
    Element 4.4.3 Communication, 52
    Element 4.4.4 Environmental Management System Documentation, 52–53
    Element 4.4.5 Document Control, 53
    Element 4.4.6 Operational Control, 54
    Element 4.4.7 Emergency Preparedness and Response, 3, 54–55
    Element 4.5 Checking and Corrective Action, 55–57
    Element 4.5.1 Monitoring and Measurement, 55
    Element 4.5.2 Nonconformance and Corrective and Preventive Action, 55–56
    Element 4.5.3 Records, 56
    Element 4.5.4 EMS Audit, 56–57
    Element 4.6 Management Review, 57
    registrars (*see* Registrars)
    registration (*see* Registration)
ISO 14020 Environmental Labeling, 160–161
ISO 14021 Environmental Labeling, 161
ISO 14022 Environmental Labeling, 161
ISO 14023 Environmental Labeling, 161
ISO 14024 Environmental Labeling, 161

Johnson, Scott David, 169, 170, 171

Klassen, Robert David, 171–172, 174, 175
Kok, Andrew, 87, 88
Koretsky, Joseph, 60, 105
Kraft, Michael E., 90

Labeling, environmental, 160–163
Landfills, 80
Legal issues, 44–48, 87–101
    environmental regulation and economic development issues, 95–96
    EPA's role in environmental regulation, 89–95

Legal issues (*Cont.*):
  public involvement in environmental policy, 97–101
Legislation
  General Agreement on Tariffs and Trade (GATT), 2
  major environmental, 30
  Malcolm Baldridge National Quality Improvement Act of 1987, 118
  Natural Resources and Environmental Protections Act, 29–30
  Resource Conservation and Recovery Act (RCRA), 31
Little, Arthur D., 158

Major nonconformity, 12, 19
Malcolm Baldrige National Quality Improvement Act of 1987, 118
Management
  environmental, 169–172
  environmental system of (*see* Environmental Management System)
  review of, 57
  Total Quality Environmental Management (TQEM), 116–122
McKiel, Mary, 94
Mendenhall, Jack, 79
Minor nonconformity, 12, 19

National Audobon Society, 98, 100
National Resources Defense Council, 100
Natural resources, sustainable use of, 33 (*see also* Conservation)
Natural Resources and Environmental Protections Act, 29–30
NCR, 12–13
Noise pollution, 80
Nonconformity report (NCR), 12–13

Observation, 12
Oil spills, Exxon Valdez, 1–2, 5, 97

Personnel, needed for overseeing ISO 14000, 128–129
Pollution, 80
Preassessment audit, 15

Preregistration audit, 15

RAB, 63
Recordkeeping, 56, 168
  EMS, 131–132
Recycling, 80
Registrar Accreditation Board (RAB), 63
Registrars, 59–67
  accreditation of, 63–64
  auditing philosophy of, 60–62
  credibility of, 59
  single-auditor certification programs for, 65–67
  steps to take before hiring, 62–63
Registration, 69–72
  advantages and benefits of, 70, 157–158
  advertising, 159–164
  competitive edge, 158–159
  costs involved in, 70–72, 172–175
  example using SGS-Thompson Microelectronics Inc., 73–82
  hypothetical example of, 122–133
  implementation steps toward, 167–169
  measuring your environmental costs for, 133–135
Registration audits, 14–20, 110
  description of, 14–15
  document review, 15
  preassessment period, 15
  preparation for, 15–16
Regulations, 79
Resource Conservation and Recovery Act (RCRA), 31
RvA, 22, 26–27, 63–64

Sandler, Blair Witten, 172
SEC, 28
Securities and Exchange Commission (SEC), 28
Seipke, Denise, 176
SGS-Thompson Microelectronics Inc., 73–82
Sierra Club, 98, 99
Social responsibility, corporate, 165–166
Stroup, Richard, 95
Suppliers, 103–106
Surveillance audits, 83–85
  focus of, 84
  frequency of, 83
  stakeholder compliant files and, 84–85

Index

Total Quality Environmental Management (TQEM), 116–122
Training, 51–52, 65–67, 107–116
  approaches to, 114–116
  choosing employees for, 113
  conducting environmental, 119
  importance of, 107–116
  internal auditor, 109–112
  ISO 14000 implementation, 112–113
  objectives of, 114
  overview of 2-day course in ISO 14000, 109
  overview of 5-day auditor course in ISO 14000, 111
  screening trainers, 114–115
Transactional audits, 31

U.S. Coast Guard, revised shipping guidelines, 2–4
United Kingdom Accreditation Service (UKAS), 18, 64

van Erp, Casper, 64
Vig, Norman J., 90

Waste disposal, reducing, 33, 81
Water conservation, 80
Water emissions, 80
Webber, Richard, 88
Weiss, Norman L., 89
Wilderness Society, 99

About the Author

Perry Johnson (Southfield, Michigan) is the founder of Perry Johnson, Inc. (PJI), the world's largest ISO 9000 training organization, and Perry Johnson Registrars, which provides fully accredited ISO certification services. He has spoken extensively on quality training and ISO certification at seminars throughout the world, and is an internationally recognized expert in these areas. He is also the author of McGraw-Hills's successful *ISO 9000: Meeting the New International Standards* and the forthcoming *ISO 9000 Yearbook*. PJI will be the first ISO 14000 accredited lead assessor and registrar in the world.